# MIAMI's GREAT HURRICANE:
## September 18, 1926

Karen Dustman

Grateful thanks to Miami historian Seth Bramson for his kind help, gentle corrections, and unstinting support. Any errors that may remain are solely my own.

The assistance of the HistoryMiami Museum has also been greatly appreciated, including research help during my visits to their Archives (a wonderful source of original accounts and photographs of the 1926 hurricane). Special thanks to archivist Ashley Trujillo and archives associate Jeremy Salloum for their kind assistance with permissions for materials and images for this volume.

*Dedicated to the staunch souls who
survived the 1926 hurricane,
and to the memory of the
many unfortunates
who didn't.*

**Clairitage Press**

First Edition

Copyright 2025 - Karen Dustman
All Rights Reserved

# Introduction

A strange, unsettled feeling settled over Miami on the afternoon of September 17, 1926. Scudding clouds gave way to high winds. Those winds soon turned to a gale. And during the wee, dark hours of Saturday, September 18th, the gale morphed into a terrifying hurricane.

That fateful Saturday would bring fear, loss, and, in some cases, injury to thousands in and around Miami. For hundreds more, it would mark their final day on earth.

My father was ten years old on that frightening September Saturday. With his mother, grandmother, and two younger siblings, he was living at Arch Creek (today's North Miami), in a two-story wooden house his father had built on 16th Avenue, near 125th Street and Old Dixie Highway. The FEC train depot sat just west of their home.

There was no adult male in the household; Dad's father had left his wife and six children five years earlier. The three oldest siblings were no longer at home, either. A sister, married just nine days earlier at the tender age of sixteen, had moved into a garage apartment at Sunny Isles with her new husband. Dad's two older brothers, ages fourteen and not-quite-eighteen, had a place of their own.

Of the day before the hurricane, he would later recall: "I remember that the air felt funny. I could sense that something was wrong. But we had no real warning until it started to blow."

And blow it did. Around midnight, gale-force winds began battering the house. And by two or three a.m., the greatest hurricane of the day to strike Miami had put in its appeareance – a Category 4 monster.

Electric service failed, and the world was plunged into darkness. By candles or lantern, Dad's mother gathered her three children and stood them in the fireplace at first for protection. But soon the waters began to rise. When the floodwater reached the first landing of their stairs, she realized better shelter was needed. She herded her children through wind, driving rain, and flooded streets to a brick warehouse a short distance away. And still the water kept rising.

*The family home at Arch Creek. The family fled when the water topped the first stair. Flood water would eventually reach about three feet high inside the home.*

Eventually a boat arrived and ferried them to higher ground. The family found shelter in an unfinished hotel-apartment building a few blocks southwest of their home,* where they rode out the rest of the hurricane with other refugees.

My dad remembered worrying terribly about his newly-married sister at the beach. They would later learn that she and her husband had been rescued. Many others at Miami Beach were not so lucky.

*The concrete block wall of a warehouse, collapsed by the storm. Official FEC Railway Photo. (Dustman Collection).*

Once the winds finally abated, the family returned to find their home a wreck. Although the house sat about three miles from the edge of the Atlantic, water had risen high enough to cover the keyboard of their Philpitt piano. The wall of the brick warehouse where they'd briefly taken shelter was completely collapsed. Had the family not moved, they could well have been buried beneath the rubble.

The family remained in the unfinished apartment building for a week or two. The Red Cross helped make repairs to their home, and eventually they were able to move back in. But not everything went back to normal. The family almost certainly lost furniture, personal effects, and household items to the floodwaters. What my dad remembered most about the damage, though, was that all of the ivories came off the piano keys. A budding pianist, he simply continued to play the instrument without them. But Dad never forgot that terrifying experience. Even as an adult, he'd become agitated and unsettled whenever a big storm moved through.

Miami, too, didn't bounce back completely from that terrifying day. Though tourists returned and much of the worst damage was swiftly repaired, the storm dealt Miami's once-booming real estate market a harsh blow. Just three years later, the city would suffer another gut punch from the stock market crash of 1929 and the onset of the Great Depression.

The Great Hurricane left tremendous damage in its wake, both human and economic. But it also left a wealth of stories. Tales of courage. Tales of survival. Tales of resilience.

My father's own story became part of our family lore, and his memory of surviving the hurricane always fascinated me. About 1990, I started collecting photographs, postcards and other memorabilia about Miami's "Great Hurricane." I made pilgrimages to the Historical Museum of Southern Florida, home to a treasure trove of photos and first-person accounts of the storm. And I began dreaming about this book.

As I write these words, one hundred years will soon have gone by since that tragic September day. My dad departed this earthly realm for a better one in 2001, and there's almost certainly nobody left alive who actually remembers that storm. But my father's memories and those of many others live on in their letters, their photographs, and passed-down stories.

Here, then, is the fearful and fascinating history of Miami's Great Hurricane – dedicated to those who survived it, and the many unfortunates who didn't.

\* *The unfinished apartment/hotel where Dad and his family took shelter during the hurricane was near a pseudo-Spanish village known as Pueblo Feliz. This sprawling, multi-block tourist attraction stretched east from the FEC tracks to today's Biscayne Boulevard, and south from NE 125th Street to roughly today's NE 121st Street. The site included a performance hall known as Teatro de Alegria. In later years, this large property would become the Thunderbird Studio, and later still, the Ivan Tors Studio. Many thanks to Miami historian Seth Bramson for his help in identifying the location.*

*Pueblo Feliz and the Teatro de Alegria (from a postcard circa 1926).*

*Etching of 'Bay Buisquine' in 1871. (Harper's Monthly).*

# CHAPTER 1:
# Before the Storm

*Ahh,* Florida. When Ponce de Leon first set foot on its northeastern coast in April 1513, he dubbed it *La Florida,* or "Place of Flowers." The pretty name stuck.

Fast-forward another three hundred years, and Florida left Spanish and (briefly) British control to become the 27th state in the Union on March 3, 1845. The freshly-minted state's population was somewhere north of 66,000 at the time (the exact total remains a mystery, since two counties, Dade and St. Lucie, didn't participate.) Nearly half of those inhabitants, sadly, were slaves.

Today 66,000 may not sound like much. But it actually reflected astonishing growth; the population had roughly *doubled* since 1830, when just 34,730 souls had inhabited the entirety of what would soon become Florida.

By 1880, less than four decades after Florida became a state, its population quadrupled. What once had been sparsely-settled outposts and a few large ranches and plantations began filling in. New settlers arrived, fresh towns sprang up, and coastal swamps were drained. Florida began to reimagine itself as a resort destination instead of simply agricultural land.

George M. Barbour trekked the length and breadth of Florida that same year (1880), publishing his observations in a volume titled *Florida for Tourists, Invalids and Settlers*. Much of what he wrote was optimistic at best, and mere puffery at worst. A large portion of the state is "never" visited by killing frosts, he blithely told readers for example. Due care in selecting land plus "cleanly habits" and "the right sort of house" would ensure settlers need not fear of malaria. As for nuisances like wood ticks and chiggers ("red-bugs," he called them), a liberal application of ammonia, camphor, or vaseline to the ankles was the simple remedy.

The one species of predator that Barbour did caution readers against was the "land-shark," one or more of which infested "almost every locality," he confessed. But these could be easily identified by certain primary traits: offers of "the greatest bargain to be had in the State," for example, and a suggested price "about twice as much as the property could actually be bought for." Hasty bargains should be avoided, he counseled, and "two or three different parties" should be consulted on the merits of any property before buying.

Even in 1880, tourism was making inroads, as Barbour's narrative shows. "Each season the army of tourists to Florida is increasing," he reported, "and the farther south they can get the better they like it." But transportation remained the single biggest obstacle. Though the South Florida Railroad was slowly pushing its way down the state's western coastline, both Key West and the future site of Miami could only be reached by steamship. Fort Myers remained a largely-isolated community of about 200 souls, most of them raising cattle.

Tellingly, Miami received no mention at all in Barbour's lengthy treatise. The city itself did not yet exist, of course. But even old Fort Dallas escaped Barbour's attention. Built at the site where Miami would eventually flourish, Fort Dallas had been constructed in 1836 by the U.S. military during the Seminole Wars. Its somewhat vague mission: to prevent the Indians from improving their finances by trading with Cuba and the West Indies, and to generally "harass the

---

*George M. Barbour, *Florida for Tourists, Invalids, and Settlers* (D. Appleton and Co., 1882).

*Mouth of the Miami River in 1871, showing remains of old Fort Dallas (Harper's Monthly).*

*"The Hurricane" (Harper's Monthly, 1871). Even in South Florida's early days, visitors and settlers knew about hurricanes, like this illustration of one striking Key West. But that didn't deter growth.*

enemy." Although the military had abandoned the fort in 1857, its buildings were pressed into service briefly as a hospital during the Civil War. An accidental fire in 1872 had wiped out all but two of the coquina-stone structures that remained. Little wonder, then, that Barbour didn't mention the site in his 1880 overview of the state.

It would be another decade before Julia Tuttle would arrive in 1888 and, in 1891, convert the Fort's two surviving buildings to her family home. With neighbors Mary and William Brickell, Tuttle championed the primitive outpost's potential, beseeching Henry Flagler to extend his railway 65 more miles.

Flagler's ambitious rail project had grown under various names and through incremental extensions since the mid-1880s. Beginning with his grand Ponce de Leon Hotel in St. Augustine, Flagler had assembled a chain of hotels and an empire of rail lines running along Florida's east coast, and by 1894, his railway line had reached today's West Palm Beach.

Tuttle and the Brickells had made Flagler an enticing offer: vast tracts of land both north and south of the Miami River, plus a spectacular site beside the river for yet another hotel in his chain. Flagler wasn't immediately convinced. But the terrible freezes of December 1894 and January/February 1895 finally seized his attention.

Flagler dispatched a pair of agents to investigate whether Miami had, as represented, remained unscathed by the freeze. Convinced by his agents that the allegation was true, Flagler finally said yes. On September 7, 1895, the name of his railroad was officially changed to Florida East Coast Railway, a name it would keep for more than 125 years. And his engineers and track-layers kept going.

Miami officially welcomed its first FEC passenger train in April, 1896. An excursion train carrying tourists from the north swept into town on May 15. And on New Year's Eve, the southernmost gem of Flagler's hotel chain officially opened for business: the magnificent Royal Palm Hotel.

Miami was just a small settlement of 300 souls when the first train arrived. But with a fast, economical connection to the outside world now secured, its future was assured. Hardly a coincidence, then, that the town was formally incorporated in July 1896, the same year the railroad arrived.

With its new accessibility, the town blossomed. And there was more to come. Physical work began in 1905 to extend the rail line another 126 miles from Homestead to Key West. Known as the Florida

*Bathers at Miami Beach (postcard folder circa 1923).*

Oversea Railway, this ambitious project was considered the greatest engineering feat in U.S. history at the time, and would eventually cost of over $50 million dollars. The first trains made the passage in January, 1912, though additional work would stretch into 1916. Separately, developer John S. Collins dreamed of a bridge connecting Miami with Miami Beach. Thanks to financing from Carl Fisher, the 2.5-mile Collins Bridge opened June, 1913 – the largest wooden bridge in the world for the day.

Soon, Miami and Miami Beach were booming. By 1920, Miami's population had soared to 30,000, a record 440% population growth with no slowing in sight. By 1925, the population figure had jumped again to 50,000. And counting visitors, there were twice that many every winter, according to a Curt Teich promotional brochure.

Tourists could enjoy golf, polo, and horseshoe tournaments. Casinos sprouted along Miami Beach. Gondoliers poled Venetian-style watercraft on the waters of Biscayne Bay. And of course there were the simple joys of sunshine and water at the beach.

Some 75 hotels welcomed visitors, and two hundred apartment houses offered longer stays. Postcards and photographs boasted beaches, bandshells, and brand-new boulevards.

Improvements struggled to keep up. The "crumbling" wooden Collins Bridge was purchased by new owners and replaced with the

12 Chapter 1: Before the Storm

*Postcard showing a casino at Miami Beach in 1924; a tourist snapshot at Sunny Isles in January 1925; and a baby elephant taking his 'daily plunge' in the ocean in January 1922 (Nat'l Photo Co.) (likely a publicity photo for developer Carl G. Fisher).*

*Pilot A.H. Gordon and his airplane promoting Causeway Realty Corp. in 1920. (Dustman Collection).*

four-mile-long Venetian Causeway. Construction commenced in 1921 by dredging the Bay to create four central islands. Ten fixed-span bridges were added, and the new causeway finally opened in February, 1926.

Paralleling it to the south was a new County causeway to connect Miami proper with South Beach. Construction on this new road began in 1917, and finished in early 1920. Unlike the Venetian, the County causeway was free. (It would be renamed the MacArthur Causeway in 1943 in honor of Gen. Douglas MacArthur.)

One tourist, arriving in town by bus in March, 1924, wrote home on a brightly-colored postcard folder: "There sure is a lot of wealth in Miami." Much of that wealth flowed directly into developers' hands. Many who came on vacation decided to stay, and real estate prices boomed. The "land-sharks" that Barbour had complained about forty years earlier now were endemic, each struggling to outdo the others.

Coral Gables announced the availability of new homesites in its recently-opened "Biscayne Bay Section" with a full-page newspaper ad in January 1926, touting its "Miami Riviera." Another realty company hired a pilot to entice potential buyers.

*When it opened the day after New Year's, 1926, the Meyer-Kiser Bank Building was the tallest building in Miami. (Miami Herald illustration, January 1, 1926).*

*Autos took advantage of the early morning hour on Flagler Street "before the crowd gets out" in this image from photographer W.A. Fishbaugh, circa 1925.*

*Damaged houseboats in Biscayne Bay after the July, 1926 storm. (Acme Newspictures, likely taken for the Relief Department of the NEA).*

Miami businesses, too, were booming. The *Miami Herald*'s publisher, Frank Shutts, confidently predicted on New Year's Day 1926 that that in another ten years, Miami's population would reach one million. Crowds were so thick on downtown sidewalks that the *Herald* also humorously suggested double-decker sidewalks would soon be needed for pedestrian traffic.

The city warmly embraced its astronomical growth as bragging rights. When the cornerstone was laid the following July for Dade County's new $3.5 million dollar, 27-story courthouse, buried inside were documents boasting of "an era of progress perhaps unequaled by any other county in the United States."

Weather had not been a significant concern throughout those boom years. On July 27, 1926, the tail end of a hurricane did sweep through, buffeting Miami with high winds. Seven boats were destroyed, and seawater flowed across Collins Avenue in North Beach. The Bahamas fared much worse. But all in all, Miami residents shrugged it off. "Loss is slight," proclaimed one newspaper headline.

The last serious hurricane to strike Miami had occurred in 1910. And even that one did not hit directly. Instead, it swept ashore at Cape Coral on Florida's west coast before marching north and east across the center of the state. Winds topped 100 mph and the storm surge "salted" a wide swath of farmland, ruining as much as ten percent of Florida's citrus crop. But for Miami, the damage in 1910 had been minimal.

As a result, nobody really anticipated the "big blow" that arrived sixteen years later, on September 18, 1926. Not to say that the brewing storm escaped notice entirely. They weren't naming storms, back then. But as early as September 14th, meteorologists from the National Weather Service were growing concerned about not one but *three* tropical storms that were gathering strength in the Caribbean.

An official storm warning was issued for Miami by the Weather Service on September 17, and the afternoon papers dutifully published the alert. Notices also went out by telegraph and telephone to news agencies, and telephone exchanges were notified as far away as Homestead, Dania, Hollywood, and Ft. Lauderdale. The *Miami Daily News* also took the unusual step of assigning a telephone operator to answer calls from concerned citizens who were unable to reach the Weather Service for updates. (That operator would dutifully stay at his post until the wee hours of the following morning, when phone service was lost.)

Later that same evening, as the storm approached Miami, hurricane flags were hoisted from the "warning tower" at the City docks and from the roof of the Federal building.

Some of Miami's 160,000 residents took heed, and began boarding up windows. But most who actually heard the news ignored it, anticipating high winds but nothing more. And many, many more knew nothing at all about the devastating storm that was heading their way.

*Hurricane winds batter palm trees in Lummus Park at Miami Beach. Water in the park was said to be waist-high when this photo was taken. The Casa Loma apartments are in the background. (From an unidentified postcard, Dustman Collection.)*

# CHAPTER 2:
# The Storm

Miami residents may not have consciously known that a hurricane was approaching. Still, there were subtle signs. My dad remembered feeling that something was wrong. Newspaperman Joe Reese reported a strange heaviness in the air, akin to suffocation. "Great beads of perspiration had exuded upon my brow, and I felt a clammy sickness," Reese would write later in his chronicle of the storm, a "physical and mental depression."

When the storm finally arrived, it "came with great suddenness," Reese noted. The barometer began falling about 10 p.m. on the evening of September 17th. High winds soon followed, and about midnight were "blowing a fresh gale." By 1:50 the next morning, the wind had reached 57 miles per hour. Its velocity continued to rise, clocking 115 miles an hour by five a.m.

Only a smattering of rain had accompanied the wind initially. But around 3:00 a.m. the sky opened up and it began to pour. Phone service failed in Miami Beach and Hollywood, then was lost across Miami as well. Most terrifying of all, the electricity went out all over town. Fearful residents, awakened by the noise and tumult of the storm, found themselves wrapped in inky darkness.

According to a hastily-issued "extra" edition of the *Miami Daily News*, newspapermen "crawled" across the bridge from Miami Beach to the mainland at 3 a.m. to deliver the latest reports, as rumors spread that a massive tidal wave had swept over stranded beach-dwellers. That wasn't quite true. But the reality was almost as bad. Marjory Stoneman Douglas (later the author of *The Everglades: River of Grass*), described Miami Beach during the hurricane as "isolated in a sea of raving white water."

In downtown Miami, meteorologist Richard W. Gray of the Weather Bureau had raised the red-and-black signal flag at 11:30 p.m. the evening before as the hurricane approached. Even then, he would later report, the force of the wind knocked him off his feet and "let him down not too gently." Gray had remained at his post in the three-story Federal building all through the night, despite the lights failing. About 4 a.m. he discovered that the weather station's rain gauge had been blown completely away. A portion of it would later be found on the roof of a nearby building.

The torrent of rain eased slightly as daylight approached. But at 5:30 a.m., a "deluge of water" descended. One observer likened it to the flow from a fire hose. Twenty minutes later, meteorologist Gray had to frantically re-calibrate his barometer when the instrument's needle slipped off the bottom of the chart. The adjusted reading showed a barometric pressure of 27.60, the "lowest ever recorded in the history of the U.S. Weather Bureau," according to the *Miami Daily News*.

Survivors would recall a continuous howl of wind, punctuated by intermittent crashes from falling buildings, breaking glass, and flying debris. Sirens wailed in the background as fire engines and ambulances struggled to make their way through debris-clogged streets. What little visibility there was was hampered by driving rain.

About 6:15, with sunrise still nearly an hour away, the eye of the hurricane arrived. Winds fell to just 10 mph at Miami's Weather Bureau office. Suddenly, the howling and crashing ceased.

The calm was deceptive and deadly. As author Joe Reese noted, "most of those who perished met their fate when they went out during the lull between 6 and 7 o'clock Saturday morning. . . . [T]he streets were filled with people curious to view the wreckage."

# Miami's Great Hurricane 19

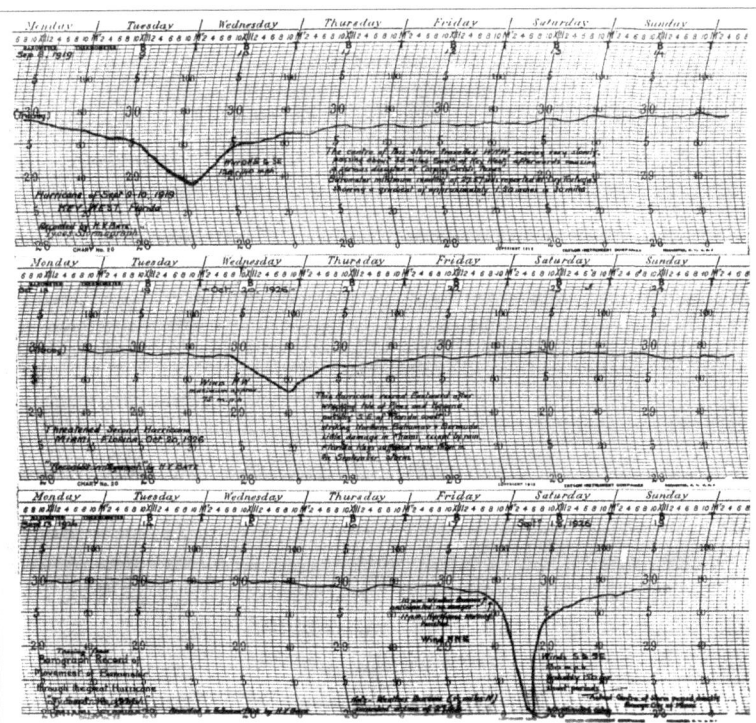

*Above, wind and water batter a seawall at Miami Beach. Below, barograph charts show barometric pressure dips. (Courtesy of HistoryMiami Archives; top image, #2000-526-1(N)).*

Chapter 2: The Storm

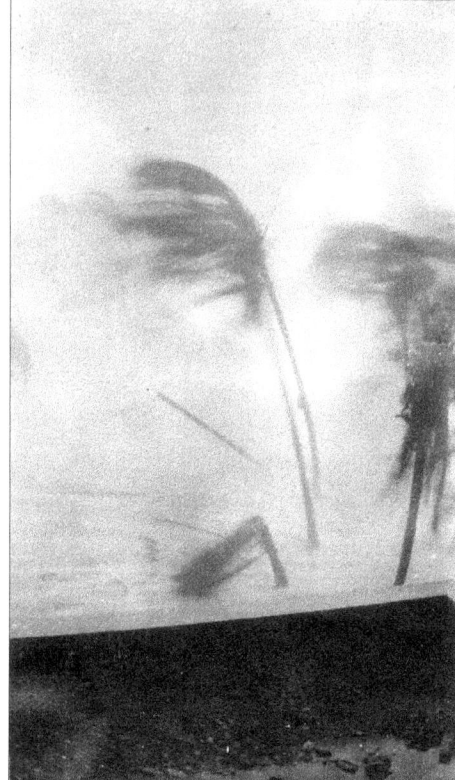

A tourist at the Pancoast Hotel on Collins Avenue in Miami Beach caught this view of the hurricane from his third floor balcony.

Below, a snapshot by the same tourist of the ruins of the Pancoast after the storm. The Pancoast was said to be the first "grand hotel" built at Miami Beach. (Dustman Collection).

*Rescuers assist a young woman on Bayshore Drive in front of the Miami News building during the storm. (American Autochrome Co.).*

The calm would last for less than an hour. By 6:47 a.m., winds were gale force again. And, according to Reese, "when the wind shifted, it was more furious than before and did greater damage."

Daybreak arrived officially at 7:07 that morning, though heavy clouds continued to obscure the sun. By 8 a.m., Miami's streets were awash under as much as two feet of water. Over on Miami Beach, the anemometer on the rooftop of the Allison Hospital blew away at 8:12 a.m., after recording 120 mph winds.

The storm tide at both Miami and Miami Beach peaked eight to nine feet above normal. Miami Beach was inundated, with the waves depositing several feet of sand on streets and inside buildings. On the mainland, the entire waterfront flooded. Water stood three to five feet deep inside hotels and homes near the bay. Boats were propelled inland and deposited two to three blocks from the bay. A sub chaser, blown clear from its moorings in Biscayne Bay, was dropped unceremoniously in Royal Palm Park.

Adding to the destruction, an 11-foot surge up the Miami River damaged boats that had sought safe anchorage there and crested the banks, flooding the northwestern part of the city. A canal at Hialeah also overflowed.

At long last, the hurricane started to ebb about two that afternoon. Residents could begin to catch their breath after eleven hours

*High winds plastered an automobile against the side of a building on First Street in downtown Miami. (American Autochrome Co.).*

*Water and debris clog the business district on Flagler Street. (American Autochrome Co.).*

*Sand, water, and uprooted vegetation block a road four blocks inland. (American Autochrome Co.).*

of torment. Frazzled survivors, many wearing bathing suits, began wading the flooded streets around 3:00 p.m., surveying the damage. Some searched anxiously for missing relatives or checked on their neighbors. Others shook their heads or cried as they inspected their homes and businesses. Some who'd had the clothes literally ripped off their backs and went in search of something to wear.

Power remained off, leaving the majority of Miami residents without electricity for cooking or lights. Downed telephone poles blocked roadways and clogged traffic. Water mains had broken, so people dipped water from the flooded streets for washing or flushing toilets.

Meteorologists would eventually classify Miami's catastrophic September 17-18, 1926 hurricane as a Category 4 – one notch shy of the severest rating. Wind speeds had reached at least 130 miles per hour, and some claimed a peak of 136 or 140 mph. The storm surge had brought the ocean more than 11 feet above mean sea level at Biscayne Boulevard. The hurricane would go down in history as the twelfth strongest and twelfth *deadliest* storm of the 20th century. All-told, it wreaked property damage across multiple states estimated at $1.8 billion in today's dollars.

Worst of all, of course, was the loss of human life. Some became casualties of flying debris or falling structures. Many caught on houseboats or pleasure craft drowned. Curiosity killed others who

24  Chapter 2: The Storm

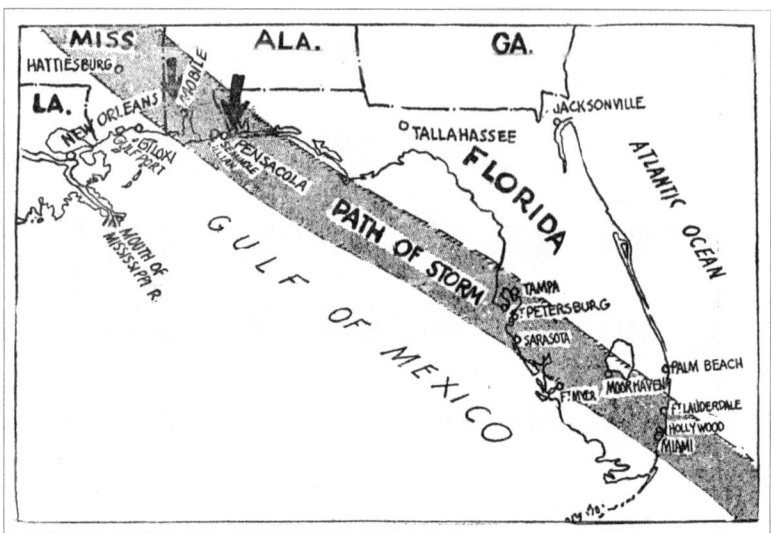

*The September 1926 hurricane hit the coast with a hundred-mile-wide swath from Key Largo almost to Palm Beach, then crossed the state and swept over the Gulf of Mexico before finally dying out in southern Louisiana. (Dustman Collection).*

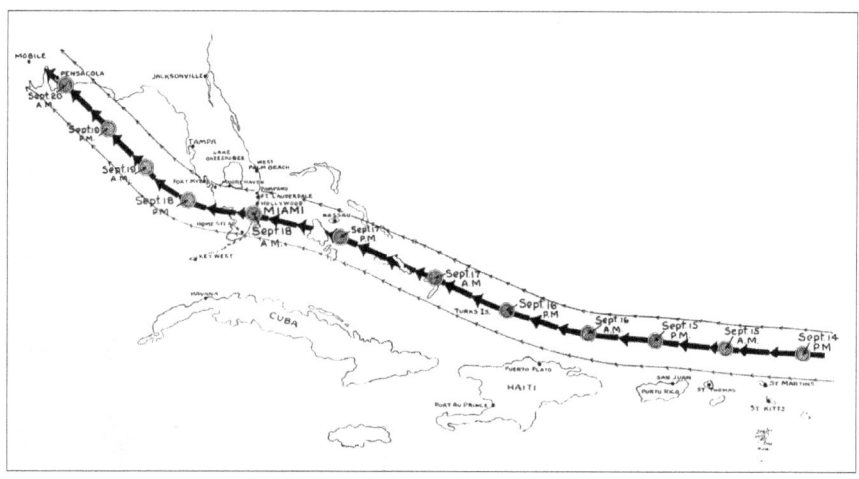

*A different sketch tracking the path of the storm from September 14 to September 20, showing its progress every twelve hours. (Florida's Great Hurricane, pub'd by Lysle E. Fesler).*

ventured too near the waves as the powerful storm swept in. Many unknowingly left shelter when the "eye" of the hurricane arrived, only to be killed when the fierce winds returned.

In that one tragic day, approximately 370 lives came to a sudden end. At least 113 were dead in Miami; another 54 in Hollywood; 22 in Hialeah; 10 in Ft. Lauderdale, and another 11 in Dania. Hardest hit of all was Moore Haven, a small town of 1200 souls, where the floodwaters from Lake Okeechobee claimed over 150 lives. The exact death total was impossible to determine; some bodies were simply swept away and never found.

Another five thousand people were injured in the storm, and as many as 50,000 found themselves suddenly homeless. Estimates of the number of homes destroyed ranged from two to twenty thousand. Property losses across Miami were pegged at $100 million.

The cost of emergency relief alone was projected to be $5 million dollars. But the entire nation rallied to support Miami in her hour of need. Roughly half that emergency total – $2.5 million – was raised from donors across the country in the first ten days after the storm.

*A man straddles a piece of furniture inside the ruins of his Hollywood home. A salvaged chair sits at right. (Dustman Collection).*

# CHAPTER 3:
# First-Person Accounts

As the hurricane raged, young and old scrambled for safety. Homes collapsed in the powerful winds, and surging waves swept treasured possessions away. Children died, and premature babies were born. Afterward, many survivors wondered how they had miraculously been spared. Others never lived to share the details of how they died.

A wide variety of amazing and frightening tales emerged in the aftermath of the terrifying storm. Some accounts were passed down only orally, through families. But others were carefully transcribed and preserved – in letters, diaries, even magazine stories.

Here are six recollections of the hurricane, told in the words of those who experienced it firsthand.

*The first – and saddest – is by Horace Howell, city marshal of Moorehaven (sometimes spelled Moore Haven), a small community by Lake Okeechobee, roughly 85 miles northwest of Miami.*

## Horace Howell

I was at work on the levee [south of Lake Okeechobee] when it crumbled that morning at 9 o'clock. A six-foot wall of water came over and I tried to reach my little home not far from the edge of the levee. My wife had prepared for just such an emergency. She had tied an inflated inner tube about each of our four children, and two about herself. Using a pair of silk stockings, she bound them all to her. When the wave came, they floated out the front door.

For an hour and a half they drifted west before the wind. Then the wind and current changed and they were carried down the bank of the canal. They floated another hour and then were swept through and over a barbed wire fence. The tube that supported my six-year-old boy, George, was punctured and he drowned, but my wife would not cut him loose. After another hour, the baby, little three-year-old Eleanor, died of strangulation. My wife couldn't keep the waves out of her face. And still they floated on.

For two hours more my wife swam with the four children, two of them dead. Realizing that her strength was almost gone, she finally freed the two drowned children and battled on to save the two that remained alive. Shortly afterward the oldest boy, Oliver, was torn away from my wife, and carried off in another direction. A cross-tie floated by and he climbed upon it. As the current carried him on, the cross-tie floated by the body of his baby sister, Eleanor. Oliver

*Burying the dead at Moore Haven, after the storm. (Lysle E. Fesler, Pub'r).*

reached down and raised the baby's head. Seeing she was dead, he let her go and she floated out of sight, half-supported by the inner tube.

About 4 o'clock in the afternoon a current carried my wife to the home of a neighbor named Steers. She and 7-year-old Laverne, the only child remaining with her, were taken in and made as comfortable as possible. That night, Oliver, who had floated to a high spot on one of the canal banks, waded back after the water had gone down, and found his mother.

It's mighty hard to lose almost half of your little family, but I'm mighty proud of that wife of mine.*

---

*Joe Hugh Reese was a well-known newspaperman in both Florida and Georgia. Born October 23, 1876 in Eatonton, Georgia, Reese worked as a reporter for the Savannah Morning News and an editor for the Orlando Reporter-Star. His booklet "Florida's Great Hurricane" chronicled both the hurricane and his personal experience during the storm.*

## Joe Hugh Reese

I had lived in the northwest section of Miami, at Sixth Avenue and 43rd Street, since April [1926] and had not become acquainted with my neighbors until the hurricane huddled us together.

When I awoke at dawn of Friday, September 17, it was with a feeling akin to suffocation and I felt a clammy sickness. I went into the kitchen to brew my morning coffee. Soon [afterward] appeared Catherine Kelly, our faithful housekeeper, who had come to us from Chicago two years ago.

"Catherine," said I, "have you ever seen a hurricane?" She replied in the negative.

"Then you are soon to see one, for this certainly is hurricane weather," said I. My words were prophetic, though at the time I had no sense of divination. My body was my barometer. All during the day I suffered physical and mental depression. In the afternoon the

---

*Horace H. Howell's story was preserved in Joe Reese's booklet, "Florida's Great Hurricane." It also was mentioned briefly in the Orlando Sentinel (September 24, 1926), and in a newspaper story on the 50-year anniversary of the hurricane (Ft. Myers News-Press, September 18, 1966).

newspapers carried storm warnings, and unusually early, the streets were crowded with motor cars scurrying homeward.

Upon my arrival at home Friday afternoon I informed my wife of the impending storm, and at once she suggested that we move the children's beds in from the back sleeping porch. I opposed the idea because it seemed needless to take such a precaution until the evidence of its necessity became more urgently apparent; nevertheless when bedtime came, we put two of the children who usually slept on the porch in my wife's bed, and I took the third [child] in bed with me on the front porch. The other [three] usually slept in the body of the house.

Soon after we retired, a fresh wind blew up and relieved the humidity of the day, but it was only pleasant and did not become alarming in bluster or velocity until the early morning. I did not know just what time it was, but as the rain began to blow in through the screen, I moved my bed from the porch into the living room, and turned on the light to guide me to the back room to see how the children were faring. The windows were open, but at that time the rain had not begun to beat through, and I returned to the living room and went back to bed. It was not to sleep, however.

The wind and rain increased in intensity, and in a short time I got up again. By this time the wind was whipping the awnings in a lively fashion and their metal frames were creaking and shrieking in a most alarming medley, which made sleep impossible. Again I punched on the light, but had hardly done so when the bulbs faded out in a dull red glow, denoting that the current had failed. My wife appeared and asked if I had any matches, and told me I would find candles in the kitchen cabinet.

Matches! With all my premonition of the storm, I had not thought of matches. But if I am improvident by nature, I may lay claim to a fairly good memory. I remembered that I had seen a packet containing three paper-stemmed matches on a chiffonier. I groped in the darkness and found them. Yes there were three, and there had been several days of humid weather. What if the matches would not ignite?

Fortunately the first match was successful, and then I misplaced the others and did not find them until after the storm. After making a light I distinguished here and there through the night and blinding rain dim splotches of light which indicated that my neighbors [too] had provided themselves with the primitive candle with which our forefathers were familiar. Despite the progress of the age, it seems inevitable that we must resort in emergency to primitive conveniences. Just across the street such a flicker was discernible, and I knew that Harry Goldstein, my friend of many years, was keeping a lonely vigil against the storm, for Harry is a bachelor and lives alone in a garage apartment which he had built only a few months ago. His domicile was a sad wreck, but he escaped unhurt, and he told me the next day that he had expected every minute to feel the house give way. I told him I thought of him during the night, and would have asked him to share our anxiety if this had been possible. He said he would have deserted his quarters had he dared go out. This fairly stated the predicament of all. They were frightened where they were, but feared to venture out.

It is wonderful what a difference the light of day makes in one's feeling of security. It is [unlikely] that anyone slept through that horrific night, except children. Our Bobby slept, though drenched. Everybody waited for the dawn, and when it came there came with it a feeling of relief, though the most severe part of the hurricane was after 7 o'clock Saturday morning.

It was like parts one and two of a grand opera program. Wagner probably gained some of his inspiration from a hurricane. Surely there was weird music in the rhythm of the storm. It was grand opera of the grandest and most awful character. At the height of its fury I played the Victrola to keep the children quiescent, but the instrument could hardly be heard above the din of the wind. Bobby prompted me when, at the conclusion of a record, I failed to change it. The child was more composed than I.

Going back into the back room shortly after my first visit, I found it leaking like a sieve. The water was pouring in upon the floor and beds. I woke Felix, our eldest child, who was sleeping soundly, and got him up, and rescued the two little girls, Mary and Jeanne. The water was streaming in upon them, and I took them both in my arms into the living room. My wife had moved baby Elise in her basket into the northwest corner of the living room, and Catherine had come in from her room, bringing our third son, Millard, with her. Here, between the fireplace and an old-fashioned davenport, the little flock hovered as the wind charged and roared and made the night hideous.

Our house is of concrete and stucco construction, having been built by a North Carolina physician as a winter home for his family. I found comfort in this reflection, for the storm was developing such fury that I began to realize it would be fatal to the cheap and carelessly constructed houses that had gone up during the boom period.

Later, when the wind shifted and the house began to quiver and the ceiling to undulate in billows, as a bedspread does when shaken by an energetic housekeeper, I realized that even the strongest construction was being subjected to a severe test.

In the meantime dawn had broken and I attempted to open the kitchen door upon the back porch to reach the refrigerator. The children were fretting from hunger. Several times I threw my whole weight and strength against the door but could not budget it, so strong was the force of the gale upon it. Eventually there was a lull and I got the door open, but wreckage blocked the passage from the kitchen to ice box. The entire screen siding, with awnings attached, had blown across the children's beds, which also had been displaced, along with two chiffoniers, in which the children's clothing was kept.

The whole house was flooded and everybody was wet to the skin. Rugs, bedclothing, furniture and everything else was in wet ruin. But just then a bottle of milk in the refrigerator seemed about the most important thing in the world.

At last I surmounted the ruins and retrieved the milk. This stayed the hunger of the little ones for awhile, for all except Millard, who indignantly refused it because it was cold – he is accustomed to taking his warm, but as the electric current was off and we were dependent on it for cooking, it was impossible to cater to the young gentleman's wishes.

The lull was of brief duration. I had begun to think of trying to get out when my wife reminded me that this was a hurricane, and so far the wind had been blowing from the northeast only and was due to change shortly. She had scarcely spoken before the shift was apparent and the real bombardment began.

The battering started about seven o'clock and continued with unabated fury for several hours. It was shortly after eleven when all of us were huddled together in Catherine's room, having left the living room for fear it would crush in, that I heard a thunderous hammering upon the kitchen door, accompanied by W.E. Sutherland's sonorous voice: "Hey there! Can I do anything to help you?"

I opened the door. There stood Sutherland in his bathing suit, water dripping from his grey hair and bronzed face, but his eyes gleaming with courage and kindness. One by one, smothered in bathrobes and other protective coverings, the children were taken to his home, which had escaped with minor damage, and our own water-logged ark was forsaken. Not only did my good neighbor take in my forlorn brood, but he and his excellent wife provided shelter and culinary accommodations for several other families. They had an oil stove and the lack of electricity was no inconvenience to them in this respect. My children became their especial charge and care. For ten mortal days and nights they humored their whims, and not only that, they displayed the greatest diplomacy and tact in their unaccustomed dealings with them. [D]isaster and misfortune make the whole world kin!*

---

*Dr. George E. Woollard's Coral Gables office was in a swanky building opposite Roney Plaza, overlooking the ocean. Its lavish furnishings included two hand-embroidered Japanese panels, Oriental chairs, desks, steel filing cabinets, bookcases, and a buffet. A separate solarium held a new Steinway piano. Ironically, Woollard had just finished moving his belongings into a two-room apartment adjoining his office the day before the hurricane.*

### Dr. George E. Woollard

I was seated in my private office that eventful Friday night after dinner, reading the paper, when the phone rang. One of the vice presidents of the Coral Gables Corporation advised me of the predicted storm. This was about 7:30 o'clock and was just barely dark. I went out and bought a paper, and upon reading it, called in my night watchman and told him to make sure all doors and windows were securely locked and barred, and went with him to supervise [the] same. Then I went back to my reading, confident that at the most it would only be about as bad as the flurry we had in July, which although it deposited several inches of sand in the office and wetted the rugs badly, did no serious or permanent damage.

About 10 o'clock the wind was getting pretty high, and I went out and turned [my] big twin-six Packard coupe around, backing it against the wind, putting it in reverse and setting the emergency brake as tight as possible. The wind at this time was so strong as to cause the sand to cut into one's face like needles. I then went back to retire, but lay down with my clothes on.

I noticed that the ocean was rising rapidly and the water already was coming under the doors. It was only a short time before I noticed

---

*From Joe Hugh Reese, "Florida's Great Hurricane" (Lysle E. Fesler, publisher, 1926).

*Sand-filled Roney Plaza after the storm. (Dustman Collection).*

the rugs were being lifted and, putting my hand down, felt the water to be about twelve inches deep, and decided therefore to change into my bathing suit. [I threw] my clothes containing upwards of $1,260 on the back of a chair. My large ring and wrist watch I took off and locked in my desk drawer. This must have been about midnight. I have no record of time for the next two days.

About this time the lights went out and, finding matches, I lit a candle which was in one of the wrought iron candelabra. The water was still gradually rising, and in the outer office I could hear huge waves swishing in great torrents when all of a sudden I heard an awful crash.

Opening the door to the outer office, I saw that the front doors had given way and huge waves were coming in, carrying out furniture on their return. Foolishly it seems to me now, I rushed over and tried to close them and hold them. The noise of the water was so great that nothing else could be heard. It was all I could do to retain an upright position, even by holding onto the walls. At this time the remaining windows and doors crashed in at intervals, until all were gone on the ocean side. I was bewildered and did not know which way to turn. I was just about to return to my private office when a great wave came in from two directions – one from the ocean side, and the other from Twenty-third Street, and a big bundle rolled up at my feet and a light flashed.

I helped this bundle scramble to its feet and it turned out to be a policeman who [had been] on duty outside the Roney Plaza Hotel.

He told me that he [had been] trying to make his way along when a huge wave picked him up and deposited him right at my feet.

We made our way back to the lounge, the northern window of which was still intact. We sat in armchairs for a while, when suddenly this one remaining window crashed and I could feel the floor trembling. The water rose so high that the chairs on which we were sitting were floating, and [when] I felt the floor quiver, I grabbed the policeman's wrist and bellowed to him to follow me outside.

We could see trees going by uprooted, huge coconut palms as well as roofs of houses. I knew not where we were going, but it must have been Divine Providence that directed me to the only room the door of which still held as I rushed through, still leading the policeman. I reached back and pulled the huge pecky cypress door closed behind me to prevent the waves coming in.

The policeman now held his torch [flashlight] in his helmet. As the water was up to our waists, we had to climb upon the table, where we sat until it rose so high that we had to stand. Minutes now seemed hours, and although we scarcely spoke a word, the knowledge that we were not absolutely alone was some comfort to both of us.

As the water rose to a dangerous height, he suggested that we had better get out. I asked him how. He suggested the door. I was, by this time, agreeable to try anything once, but on making our way to the door we found the sand and debris had piled up above the catch-sash of the windows, to say nothing of the fact that a raging torrent was racing by the windows carrying everything in its wake, including a huge piano, a relic from the house of the Emperor of Austria, which went floating by on a wave at a terrific rate as though it had been a feather. Escape was cut off.

At this time we saw two huge roofs crash to the ground right outside our windows. These had acted as wooden awnings for [my] office and the second floor of the casino above, and were completely demolished. A few moments later, we saw the Roman Pools split completely in two, releasing 500,000 gallons of water in one deluge, having the same effect as a dam bursting and carrying everything before it.

The water rose rapidly, and the most welcome sight anyone ever saw was dawn. We had been forced to stand on the table and hang onto the chandelier, and from this point the water rose steadily until it was at our chins, with our heads touching the ceiling. We could go no higher, and I told the policeman that if a lull came for only a few

Miami's Great Hurricane 35

*One of the Roman Pools before the storm. (Dustman Collection).*

minutes, I was going to make a break for it. How, I had no idea. But he protested it would be suicidal, as roofs were going by on the wind, and four were lying in what remained of the Roman Pools, and the debris was still racing by the windows at a terrific rate.

I stepped off the table and swam under a door to a small closet adjoining this room and, on coming to the top, was overjoyed to see the window there had not become blocked by debris. If I could smash it I could get out and possibly make my escape by getting back into the solarium. I swam back and so reported to him. He could not see it, but finally we agreed that I was to go and that if I should be swept away, he was to attempt to recover me.

I took his pistol and, swimming under after shaking hands with him, I reached the window sill. After trying unsuccessfully twice, [I] finally smashed the window and then the screen, dropped the revolver, and stepped gingerly from the window.

I had, however, barely got my right leg and arm through before I was picked up as though I had been nothing and thrown bodily back into the pile of debris and piano. I picked myself up and, clinging to the balcony above me, made my way hand over hand to the windows of the salon. The wind was now coming from the south at terrific force, making it difficult for me to retain my hold. Finally tearing me loose altogether, it hurled me through the windows and up against a marble fountain in the center of the solarium. I made my way, swimming and crawling, to the mantelpiece where I rested.

The tide was now receding and the waters in the office also were going with it very fast. Finally I swam and walked where the sand was exceptionally high to the door of the room in which the policeman was still a prisoner. My hand hit something metallic which turned out to be a frying pan, with which I gradually dug him out. The wind was still raging so that we could not stand erect. We made our way to the mantel.

At noon on Saturday we were rescued and taken out, after having been in the water for over ten hours. I had nothing but a bathing suit on, [and] even this was partly torn off, and for the next four weeks I was forced to live in a borrowed one. The office was stripped bare of all its beautiful furnishings and decorations. Even the piano was smashed to pieces and was found a half block from where it formerly stood. Sand in some places in the inner office was over six feet deep, and was nowhere less than two feet deep. Floors of solid concrete and tile had collapsed entirely. Some weighing over five tons were carried as far as 50 feet away. The huge front entrance arch of solid concrete, standing 40 feet high, had fallen completely into the street. The beach, which was formerly level with my office, was now five feet beneath it.

The twin-six Packard mentioned in the first part of this recital was turned upside down and carried a half-block away, completely buried in sand, and was utterly demolished. Two of my typewriter desks, with typewriters intact except for saltwater rust, were found on the middle of the golf course approximately 1-1/2 miles from the office. Pieces of some of our antiques were found several blocks away, and the steel filing cabinets were undoubtedly carried out to sea on the return of the waves. Of everything formerly contained in this office we have salvaged only nine rugs.

Drinking water was unobtainable, and the first mouthful of food I had was Saturday night, when I was given a plate of soup and a biscuit. Everything I had was literally swept away, leaving me nothing in the world but a bathing suit. We are going to rebuild, and carry on as before. *

---

*Recounted in Joe Reese's 1926 booklet, "Florida's Great Hurricane." Dr. George E. Woollard threw himself into medical relief work following the hurricane. Three days after the storm, he would discover that he had suffered six broken ribs during his narrow escape.

*The Everglades Hotel. (Dustman Collection).*

*Newly-minted schoolteacher Myrtle Burnett arrived in Miami the evening before the hurricane struck. She wrote home the next day to her parents and siblings in Mississippi, on stationery from the Everglades Hotel where she was staying.*

## Myrtle Burnett

Dearest homefolks,

Miss Rosser and I are resting up in the lounge room of this hotel, which overlooks Bay Biscayne. Ships, big ships, are across the way in Royal Palm Park. They were turned over and driven in by the hurricane.

Let me say as briefly as I can that the hurricane was fierce. News reports couldn't exaggerate how terrible it has been, the damage that has been done, and the horribleness of the situation. They can't exaggerate, because the most superlative adjectives are too mild to describe the disasters.

When we stopped in West Palm Beach ten minutes, Mr. and Mrs. McLelland invited us to spend the night with them, as a hurricane had been predicted for Miami. Well, of course we couldn't stay, so on we came. When we were taking our walk Friday night down Flagler Street, the storm had begun but we were ignorant of it. While I was taking my bath at 12:30 Friday night, the wind was blowing terribly. We went to sleep about one.

About two or two thirty, we were awakened by windows crashing and the like. Lights were out, water was cut off, everything was black as black, black night. Then vivid flashes of lightning would brighten the room. The wind blew at 75 to 100 miles per hour.

I kept lying on the bed awhile. Miss Rosser dressed [and] handed me my clothes. Then I dressed. We moved to my bathroom to keep away from the glass.

At last dawn came, and with it gradually came the calm. So at seven o'clock we went out to hunt breakfast and see if Miss Rosser's bus [was going] to Homestead. Well, her bus didn't go. On the way down we tried to get breakfast [but] there was not a place that could even serve coffee.

At last we passed a Greek restaurant that served coffee, toast, and iced cantaloupes. While we were there eating, the second [phase of the] hurricane came. We sat there for safety. Just about 10 o'clock the worst came. We had to dash across the street to the Cinderella ballroom for safety, and part of the ceiling of that fell in. They said that was the most substantial building because it was steel-reinforced concrete.

When I went across the street, a policeman accompanied me. I went through water to my waist, so of course my dress, shoes, and all were soaking wet. No telling what kind of a cold I'll have from it. My shoes are ruined. We got milk and cake for dinner on our way back.

After that [part of the] hurricane we went back to my room. The screens and storm shutters had been blown out and bent respectively. So we stayed there just a few minutes, then went on the streets to see the wreckage. Oh, I don't believe there is a building that hasn't been damaged. We bought an extra, [which] stated that thousands of dollars of damage had been done, but folks say millions have been ruined. The Meyer-Kiser building, valued at 2-1/2 million dollars, was struck and it has to come down. They have [already] begun taking down the remains of it. [*Ten of the 17 stories were removed.*] The Realty building has to be taken down, also.

A storm was predicted for 4 o'clock again. We rushed to the post office for safety. A huge rain came, but no hurricane came. Then we went back to the room. I was so exhausted that I fell across the bed. We slept from 5 to 8 [and] were awakened by a downpour of rain. Goodness, it seemed like a cloudburst.

Fortunately, our hotel had a slanting roof so the water did not stand and cause the roof to [fall] in as it did in lots of buildings. When we woke up the sky was clear. We went out to look for breakfast. Having just a dash of water, I washed my face. Water to drink is not available. We have to drink milk or coffee or hot Coca-colas and the like. Just think of the unsanitary conditions due to lack of water to use. Oh, you know there will be an epidemic after this.

We took a bus out to my school. While it is standing, the walls were crashed in. It was surely a pretty building formerly. I don't expect we'll have school for a month as people are living in the rooms where the walls were not crashed in.

News bulletins posted in the *Miami News* building report Hollywood and Fort Lauderdale are blown away.

The ambulance is passing by nearly every five minutes with the injured found in boats on the Bay. I sent you a telegram yesterday. It had to be cabled to Cuba, [and] from Cuba to New York, and then telegraphed to Meridian, [Mississippi]. I hope you've gotten it. I haven't gotten my trunks yet.

I should be and am glad to be alive and safe. It is fortunate that I had my room reserved and paid for. The town is under martial law.

Will write later. Don't worry.

Love, *Myrt*

[P.S.] Willie, do you and Rebecca wish you had come instead of me? Mamma, did you think of what I had said, "I might not get there to teach school?" Well, I most didn't save my life after I got here. Believe me, I've run from place to place for safety.*

---

\* *Myrtle Burnett was born in McComb, Mississippi, and raised in Meridian, Mississippi. Following the hurricane, she would teach at Robert E. Lee Junior High School and Miami Senior High. She married Dr. John Carter Branham in September, 1935, and they had one daughter, Gloria Ann Branham. After her husband's death in 1976, Myrtle moved to Fort Myers to be close to her daughter. She died there on January 15, 1989. Her letter is preserved at the Historical Museum of South Florida (folder #1992-311).*

*Despite the hurricane damage to its printing facility, the Coconut Grove Times released a typewritten issue less than a week after the hurricane.*

## Coconut Grove Times
## September 24, 1926

*Hurricane Number the First.* We hope 'tis the last we will issue under such circumstances. Our printer's plant is a wreck, so this is the best we can do this week.

REPORTS OF THE CALM THAT FOLLOWED THE STORM.

Everyone is interested in the experiences of his neighbor during the hours of danger, and the *Times* has collected as many as possible during the past hectic days.

One of the weirdest is told by the family of William Gatlow, who lived in a bayfront home in Bay Homes. Their cottage having been invaded and demolished utterly by a lighter [a flat-bottomed barge], the family moved in next door. The rising waters pursued the family to the second floor, a hole was made in the ceiling, and they sat out the storm in the attic.

Across the highway at about the same time, the roof of the Thomas Wyatt cottage in Ye Little Wood sailed away. The Wyatts, their guests the Fulmers, and Chris Witkow then took refuge beneath their automobiles, lying in pools of water. A huge light pole, complete with transformer, crashed down between two of the cars as neatly as if arranged by mathematics. No one was injured, and in the calm, the family took possession of a neighbor's house. Ye Little House was shoved half off its foundation, and the home of T. Spicer-Simpson, also in Ye Little Wood, is minus a roof, a verandah, and a studio.

Entrada was flooded and the second floor was the place to be, if any. "Dad" Trephagen hung from an electrolier in seven feet of water for four hours, after the grand piano on which he had stood was turned over under him by the swirl of the water. Mr. and Mrs. H. DeB. Justison left their home after all was over, with everything turned over and contents scattered, and floors afloat. A twelve-foot beam charged through the house lengthwise and lodged on the kitchen sink without damaging anything en route.*

---

*\*Both this rare typewritten issue and the following letter are preserved at HistoryMiami.*

*Cyrill Berning, a staff writer for the Miami Daily News, described his experiences in downtown Miami during the heart of the storm in a letter he wrote the following day to a relative, George A. Berning, on Daily News letterhead.*

### Cyrill Berning

*Sunday, September 19, 1926*

When the storm "broke" at about 2:30 a.m. yesterday, I was at the police station. Wires were shaken down, the nurses' home at Jackson Memorial Hospital (a city institution) was so damaged that the nurses were forced to the main building of the hospital, [and] the hospital power plant was disabled and the smokestack fell. The "stand-pipe" downtown buckled and then collapsed, shutting off the city's water supply.

The Meyer-Kiser bank building, a 16-story new structure and one of Miami's proudest, rocked; the steel girdings bent, and the bricks and concrete were shaken out and hurled to the street. The building will have to be entirely rebuilt. The Park Theater building collapsed. The Olympia Theater and Cromer-Cassel department store electric signs, said to be the largest in the south, were smashed. Henrietta Towers apartment building lost its towers. Store windows were smashed downtown, and looters started stealing jewelry, clothes, etc. early. On account of it being a Jewish holiday, many merchants had no idea of coming to their stores. The looters sure did have paradise for a while. All of them got new clothes and things. Looting and food profiteering were stopped by police with harsh measures.

At 4 a.m. [during the storm] we got a call at the police station that a man was injured when his auto crashed into a pole. I started out with the ambulance with the driver and two cops. We got as far as N.E. 2nd Street and Bayshore Drive. Signs, wire, roofing and glass were blown around us. The wind was terrific.

We abandoned the ambulance (the city's new one), one by one. Two of the policemen were blown into the new Columbus Hotel. The third one was blown up the street on his stomach and stopped by clinging to the spokes of a parked auto, into which he climbed to get out of the wind. I was the last to leave. Like an angel I flew, first catching hold of a headlight of the parked auto, and then into the Leamington Hotel across the street from the Columbus.

A little while later, the cop got out of the parked car and he, too, found refuge in the Leamington Hotel, where we got something to

*The Hotel Leamington touted itself in a 1924 postcard as the "most modern hotel in Florida." (Dustman Collection).*

eat. A few minutes later the ambulance was picked off the street by the wind and turned over on its side against a building. Both the Leamington and the Columbus buildings were badly damaged, especially the Columbus, where outer walls broke off.

All day I waded into water nearly up to the knees with a policeman, who made the rounds of the city. Bricks and debris were blown from great heights. Automobiles were blown over in the streets, and those which were near the bay were blown into it. Tops of our beautiful royal palms were blown off. Other palms, like the Washingtonians, were broken off at about the middle of the trunks. Royal Palm Park looks devastated, but the old Royal Palm Hotel, known all over the country, withstood the storm's ravages very well, suffering but slight damage. The wind and rain were the most terrible I have ever been in. Many boats were washed up out of the bay.

So far, 41 bodies have been recovered. It is estimated that not more than 150 persons were killed and 1,000 injured. During the storm it was impossible for police to reach injured persons or even to obtain information of places where houses were collapsing. A city detective, J.M. Driggers, left his house with his wife and three-day-old baby just as it caved in. City buses were used to carry homeless people from the various tourist camps and tent cities to schools and other public buildings.

After the hurricane subsided about 2:30 p.m., all available autos were put into service. The old ambulance was used until the tires were punctured by the debris under the water in the streets. The new ambulance was towed to a garage from the place where it was wrecked. A block from this place, a city ladder wagon was stalled. During the storm a fireman, Leon Rooney, was blown from the rear end of the ladder wagon and was hurled against the side of the Exchange Building, suffering a severe gash in the head. He was taken to the Exchange Building and given first aid until after the storm, when he was removed to the hospital.

*Cyrill*

P.S. The *News* tower stood up all right, but a big ornament fell from the top of it, landing in the roof of the building proper. It crashed through, wrecking the Sunday editor's office and narrowly missing two reporters who were lolling about and sleeping after having worked all night. The editorial room was flooded with water.

*Tired residents began to inspect flooded streets and damaged buildings after the hurricane had passed. (Dustman Collection).*

# CHAPTER 4:
# The Immediate Aftermath

At long last, the horrific wind died down about 3 p.m. on Saturday, September 18, and residents began poking their heads out to take a look around. What they saw was both astonishing and heartbreaking.

Roofs had been plucked off buildings, and sturdy concrete block structures lay in ruins. Telephone poles canted at odd angles, or had been pulled completely out of the ground. Palm trees were uprooted or snapped in half. The Miami streets were clogged with a combination of dark, sticky mud, building debris, and soggy seaweed. Dead fish littered the waterfront and sometimes were found hundreds of yards inland. From city streets to residential lots, homes to businesses, everything was littered with shattered glass, broken building materials, and uprooted vegetation.

*Three cars buried in the rubble of a house. (Dustman Collection).*

*What's left of a home on NE 16th Street. (Dustman Collection).*

*Cars struggle through still-flooded streets immediately after the storm. (Hurricane Scenes, E.P. Wheelan, publisher).*

*One of the classic images from the storm, taken when the water had begun to recede. (Hurricane Scenes, E.P. Wheelan, publisher).*

*A deserted lunchroom with water still covering the floor in the aftermath of the storm. (Dustman Collection).*

*Workmen assessing the shoreline while stormclouds remain overhead. (Dustman Collection).*

*Survivors comb the wreckage of their home for possessions to salvage. (Dustman Collection).*

*A family left homeless by the storm managed a smile for the camera. (Dustman Collection).*

*Broken trees block a road in the storm's aftermath. (Dustman Collection).*

*On Everglades Avenue in Palm Beach, power lines fell one way while trees bent in the opposite direction as the high winds changed course. (Dustman Collection).*

*Looking west at piles of sand and debris near the Roney Plaza Hotel at 23rd Street and Collins Avenue. (Dustman Collection).*

At Miami Beach, waves had deposited three to five feet of sand in their wake. Survivor Joe Reese described a steel flagpole in front of the Miami Beach fire station as "bent into a triangle," and reported that "palms, palmettos, and weather vanes all canted toward the northwest."

There was no electricity for lights or cooking, and telephone and telegraph lines were down. Adding to the misery, fresh water was unavailable to drink or for flushing toilets.

An assortment of vessels, both large and small, had come in on the storm surge and been deposited high-and-dry as the waves receded, some as much as four or five blocks from the shoreline. These included the WT and B Co. No. 38, a large freighter that came to rest in Royal Palm Park. The five-masted Rose Mahoney, a San Francisco schooner, was unceremoniously dropped at the foot of Northeast Eighth Street by the angry sea. As many as 500 boats were reportedly lost or damaged in the storm.

Perhaps the most-photographed nautical casualty was the *Nohab*, typically dubbed the 'Kaiser's yacht' in commercial postcards and amateur photos. Built by the famous Krupp shipyard at Kiel, the ship had indeed once been owned by a former German kaiser.

*Above, mountains of sand outside a Miami Beach home. Below, a rare interior image from Miami Beach after the water receded. The homeowner noted on the reverse that the water level had reached the second floor, the roof was gone, and "almost everything else" had been carried away. (Dustman Collection).*

*Damage and destruction at Miami Beach, adjacent to the Wofford Hotel. (Verne O. Williams photo).*

*Automobiles and broken palms littering the beach. (Dustman Collection).*

*Multiple amateur photos of the wallowing Nohab were snapped after the storm was over. As the image below shows, the hulk became something of an improptu tourist attraction. (Dustman Collection).*

The ship had passed into the hands of a company planning to offer passenger service to Nassau. Before being wrecked, it had functioned as something of a tourist attraction/dinner club, allowing patrons to gawk at its luxurious silver bathtubs, once used by German royalty. After the hurricane, it became a tourist attraction in different way.

The ship itself wasn't the only casualty; Captain Ermann and four of the *Nohab*'s crew perished in the hurricane. Only the ship's engineer, lucky enough to be ashore when the hurricane struck, had survived.

*Two views of the Rose Mahoney, a five-masted schooner from San Francisco that washed up at the foot of NE Eighth Street. Note the donkey pulling a cart at left in the lower photo. (Dustman Collection).*

*Right: The WT&B Co. freighter was beached near the Royal Palm Hotel. (top: FEC Photo; middle and lower: Dustman Collection).*

*Small boats washed ashore. A portion of the Royal Palm Hotel is visible at far left. (Dustman Collection).*

*Preparing to remove a boat. The Royal Palm Hotel is at right. (Dustman Collection).*

*Foggy amateur photo of a stranded war ship. (Dustman Collection).*

*The Miami skyline on October 4, 1926, showing a damaged roof, leafless trees, and missing exterior sheathing on the tall building in rear. (FEC Photo).*

Miami's business district, too, was hard hit. "It did not seem, as I walked down Flagler Street, that a solid plate glass window had been left in the town," recalled newspaperman Joe Reese. "Nearly every storefront had been crushed in and the stocks [of merchandise] ruined or damaged."

The newly-finished 17-story Meyer-Kiser Bank Building was among the worst casualties among downtown structures. At its opening just nine months earlier, promoters had touted it as the tallest building in Miami. Survivor L.F. Reardon, observing during the gale from the nearby Ritz Hotel, claimed the Meyer-Kiser had "waved its tail like a porpoise and did a sort of Charleston." Damage to the building was extensive, and the Meyer-Kiser became famous afterward from a flurry of postcards and snapshots documenting its shattered exterior. Initial speculation was that the entire building would have to come down.

Even before the winds had completely stopped, looters emerged. Broken shop windows and empty streets proved too tempting a target, and opportunistic thieves began stealing clothes, jewelry, and

*Above, the Meyer-Kiser building as it appeared in a newspaper sketch at its debut in January, 1926, and immediately after the storm. (V. O. Williams photo). A close-up of damage, below. (FEC Photo).*

*Sentries were posted to deter looting. (Hurricane Scenes, E.P. Wheelan, publisher).*

other goods from damaged stores. The Miami police stepped in quickly to halt such callous depredation.

If there was good news, it was that many larger structures *did* survive; three-quarters of the 150 hotels in Miami, Miami Beach, and Coral Gables escaped serious damage, and 70% of apartment buildings came through with only light damage according to Mayor Edward C. Romfh.

Almost miraculously, both the Roman Catholic Church and its nearby school escaped unharmed. Quipped Reese, "Either this shows that the Catholics are in high favor with the fates, or they know how to choose their contractors."

Mayor Romfh happened to be out of town when the hurricane hit. But acting mayor James H. Gilman took action immediately, hastily convening a "Citizens' Relief Committee" the evening of September 18th to coordinate disaster response efforts.

The American Legion had already stepped forward, opening its headquarters on Bayshore Boulevard as an emergency hospital even while the storm was still raging. In the immediate aftermath, Legion members not only helped distribute food and supplies but also assisted police with keeping the peace.

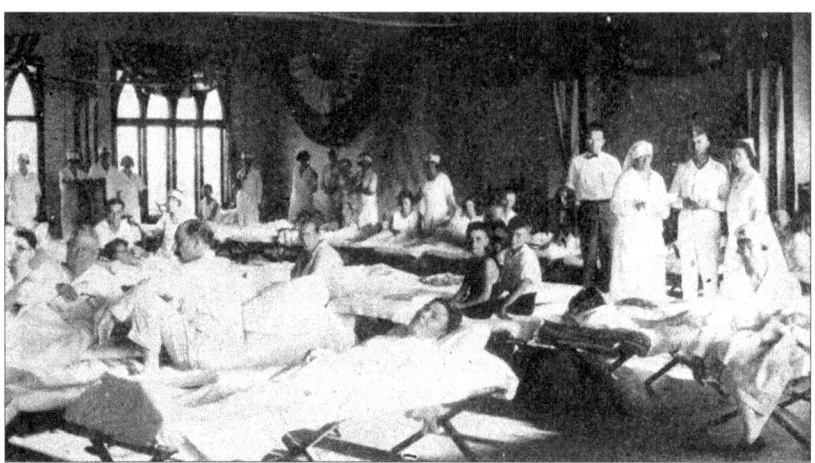

*The American Legion headquarters was turned into a temporary hospital. (Florida's Great Hurricane, Lysle E. Fesler, Pub'r).*

*The American Red Cross at Boca Raton. (FEC Photo, Dustman Collection).*

*Red Cross and relief volunteers making sandwiches for survivors. (Acme Newspictures, September 28, 1926).*

*Both the Flamingo Hotel at Miami Beach, upper left (FEC Photo), and the McAllister Hotel at Miami, at right (postcard, Dustman Collection) opened their doors to offer medical care and shelter to hurricane survivors.*

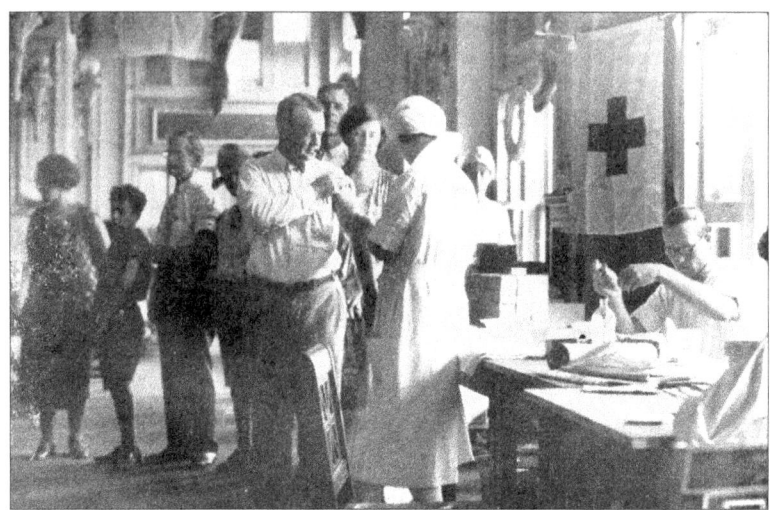

*Vaccination stations were set up by the Red Cross to provide free inoculations against typhoid fever and tetanus. (Tyler Publishing Co., A Pictorial History of the Florida Hurricane).*

*Relief workers hand out milk to children. (Acme Newspictures, September 28, 1926).*

*Crowds gather outside Miami police station, waiting for word about missing relatives. (Acme Newspictures, September 28, 1926).*

Flying debris and falling buildings produced a myriad of casualties from broken bones to fractured skulls, and local hospitals were soon swamped with victims. Hotels were hastily pressed into service as both makeshift medical facilities and emergency housing. In Miami, the McAllister and Columbus hotels opened their doors to help, while on Miami Beach, the Flamingo, Roney Plaza, William Penn, and Floridian hotels similarly offered aid. With ambulances in short supply, volunteers used their own personal vehicles to transport the injured. Schools and other public facilities, too, were turned into emergency housing or first aid facilities. And citizens opened their arms and their doors, taking in those left homeless by the storm.

Citizens soon began swarming telegraph offices, trying unsuccessfully to send messages to loved ones, only to discover the wires were down. Frenzied survivors gathered in front of the Miami police station awaiting word on the fate of missing relatives. Undertakers' parlors were filled with corpses, some with a name scrawled on a tag attached to their ankle. Lists of the dead – some identified, some still anonymous – filled the newspapers.

*Stacks of caskets for the hurricane victims at Belle Glade, Florida. (Dustman Collection).*

*Bathing suits were a logical choice in the wake of the storm. Here, men survey the flooded street outside the Miami Beach police station. (The Florida Hurricane: One Hundred Views, American Autochrome Co. (Chicago), publisher.)*

A curfew to maintain order and deter looters was swiftly imposed by Miami's Chief of Police, requiring citizens to be off the streets by 6 p.m. Able-bodied men were quickly deputized and several hundred Marines and National Guards were brought in for a short period to help patrol the streets. Although the term 'martial law' was bandied about, it never was actually imposed.

*Cigarette in hand, a man in a bathing suit stands in the doorway of Quigg's Grocery at Miami Beach. (Tyler Publishing Co., A Pictorial History of the Florida Hurricane).*

Officials did take legal steps to prevent price-gouging, however. City Manager Frank H. Wharton sternly warned retailers and food sellers they could be arrested and their inventory confiscated if they attempted to "profiteer" on the emergency.

Citizens coped as best they could. Kerosene, oil, and alcohol stoves were pressed into service for cooking, and candles became the go-to resource for lighting. At first, even clothing could be hard to come by; some hurricane victims reportedly showed up for work still clad in their bathing suits.

Horrifying rumors soon began to swirl, upsetting survivors even more. Thousands of decomposing bodies were said to have been found at Miami Beach, and thousands more were purportedly dead at Fort Lauderdale. Hysterical citizens flocked entrances to the causeways, only to be barred from crossing. With communication to the beach communities still unrestored, cold, hard facts were impossible to obtain.

But within a day after the winds abated, the Red Cross stepped in to coordinate emergency response efforts, setting up temporary hospitals and relief stations. An FEC train arrived on Sunday, September 19th, less than twenty-four hours after the hurricane had struck, bringing the first wave of relief supplies.

*Debris clogs the road at Washington Avenue and 5th Street in Miami Beach. (American Autochrome Co.).*

*A trailer arrives to remove a tugboat named the 'Escort' that had run aground near Burdine's department store. (Dustman Collection).*

*A grader and work crew clear the road of sand, likely at Miami Beach, on October 4, 1926, while a lone palm tree shows the effect of the hurricane. (FEC Photo).*

Clean-up efforts began quickly. Able-bodied men were recruited to help clear the streets of debris and to volunteer as temporary police. Prisoners were released from jail to help open the roads so relief supplies could reach nearby Hollywood, Dania, and Fort Lauderdale. And at Miami Beach, a fleet of tractors, graders, steam shovels, and an estimated 400 trucks were deployed for roughly a week to move the mountains of sand and debris the wind and water had left behind. Given the scale of the destruction, normal building permit requirements were waived for thirty days so repairs could go forward swiftly.

All public utilities remained out for roughly 24 hours after the storm had passed. But by Monday, September 20, fast-working linemen had managed to restore electric service to portions of downtown Miami. City water lines began resuming service the same day. Fearing contamination from the breaks, however, officials continued to urge residents to boil water before using it for drinking or cooking.

*Homeless children near a makeshift stove in Miami. Residents were urged to boil water, but primitive cooking conditions contributed to the outbreak of numerous fires. (Relief Department of the NEA, September 24, 1926).*

That same day, President Coolidge issued a nationwide call for relief contributions to assist the people of Miami, urging that donations be sent to the American National Red Cross. The City of Miami would issue its own appeal for funds on September 23rd, describing the once-thriving metropolis as prostrated. "Miami needs money quickly and in large amounts," it confirmed.

On Wednesday morning, September 22, a relief train organized by William Randolph Hearst and the Chicago *Herald-Examiner* steamed into town after a 38-hour run, carrying doctors, nurses, surgical equipment, milk, and water-treatment equipment. A total of ten special relief trains would eventually arrive.

By now, aid stations had been opened at various points throughout the city, offering not only medical treatment but also food and clothing to those who'd been left destitute by the storm. Vaccinations were offered against tetanus and typhoid.

Distressed residents' efforts to cope under the primitive conditions led to their own kind of danger. In the five days following the hurricane, the Miami Fire Department was forced to respond to some 1,500 unintentional blazes. "Do not light fires," the *Coconut Grove Times* sternly warned readers. "Fire will spread too rapidly to be controlled, in the debris. We don't want any big fires."

By the time an official FEC Railway photographer made it to town on October 4, the trees remained bedraggled but cars were parked beside the bandshell in Royal Palm Park – possibly workers assisting with repairs. For many, however, living conditions remained unsettled. Some, like my own family, lived in hotels until their homes could be repaired, often with the assistance of the Red Cross. Some moved in with friends or relatives. But many residents

*Above: The remains of the bandshell in Royal Palm Park on October 4, 1926, showing long lines of parked cars. (FEC Photo).*

*Below: A postcard of the bandshell in happier days. (Dustman Collection).*

*Workers make repairs to the Royal Palm Hotel on October 4, 1926. The hotel would successfully open as usual for winter visitors, but the hurricane marked a turning point in its fortunes. (FEC Photo).*

simply sheltered in the ruins of their damaged homes, or in whatever temporary housing they could cobble together.

As many as 800 hurricane survivors took advantage of the FEC Railway's generous offer to leave Miami and travel north for free. But most folks just grinned and bore the temporary hardship.

*Right, top: A man sits on the porch of his roofless house. (Dustman Collection).*

*Center: A homeowner snapped this shot of his kids and drying laundry after the storm. (Dustman Collection).*

*Bottom: A woman and her mother in their makeshift home at Ft. Lauderdale after the hurricane. (Dustman Collection).*

*The ruins of a causeway, immediately after the storm. (Dustman photo).*

*Palms line a flooded Flagler Avenue at the departure point for the ferry to Palm Beach. (The Florida Hurricane, American Autochrome Co., publisher).*

*Sign painted on a wooden door, left by a resident who abandoned Miami for Tennessee. (Courtesy HistoryMiami Archives, #1999-493-1).*

By October 9, a mere three weeks after the devastating hurricane had struck, the County Causeway connecting Miami to Miami Beach was able to reopen. On that same date, the Red Cross proudly recounted its heroic efforts to date: medical and nursing services provided to 113,200; food to 6,500; clothing to 4,650; and tents for 11,900. Relief efforts certainly weren't done. But a great deal had already been accomplished.

National Guard units remained on duty to assist with keeping order. Their numbers were being drawn down, but it would be November 2nd before the final troops left hard-hit Moore Haven.

The unnamed hurricane had been "probably the most destructive storm in the history of the United States, insofar as property loss is concerned," concluded newspaperman Joe Reese, estimating its damage at between $154 and $159 million over a 60-mile swath. The Citizens' Committee pegged the losses even higher, reporting $165 million in damage to the Miami district, of which $300,000 was for auto losses and $500,000 for plate glass. A virtual flood of insurance adjusters swept into town in the hurricane's wake.

Efforts to rehabilitate the city's image kicked into gear almost as quickly as the relief missions. "I see no reason why this city should not entertain her winter visitors [in] the coming season as comfortably as in past seasons," cheerfully proclaimed Mayor Romfh on

September 23rd. After all, he added, "The sun has been shining the past five days."

It was an attitude Miami tourists eagerly embraced. "All is calm," wrote the author of the happy snapshot below, taken by a visitor just weeks after the hurricane had passed.

*Candid shot of bathers at Miami Beach. (Dustman Collection).*

*This channel buoy, cast ashore by the storm, made a great tourist photo backdrop. (Dustman Collection).*

*A woman in a flapper hat crosses the road as repair efforts are underway. (Dustman Collection).*

# CHAPTER 5: Recovery & Rebuilding

For many survivors, life immediately after the storm centered on the basics: finding food, water, shelter. Meals were eaten on "many a bare floor," or served "under the cerulean skies of Florida," recalled newspaperman and hurricane survivor L.F. Reardon. But even amidst the shattered buildings and piles of debris, Miami's unbroken spirit shone through. A scattering of American flags began to wave atop the rubble in defiant celebration of the Magic City's survival.

*Resurrection. Resurgence. Resilience. Rebirth.* Those were the watchwords of the day.

Relief agencies swiftly trucked in basic commodities like milk, bread, and water, and began distributing them. Other much-needed supplies like roofing material and ice proved harder to come by, at least for a while. But eventually building supplies, too, were hauled in to assist residents of the damaged city.

Crews of linemen worked tirelessly, restoring electricity street by street and block by block. Within a week after the storm clouds had swept away, telegrams could again be dispatched to worried relatives and telephone service had been successfully restored to three thousand customers.

Sheriff Chase ordered all able-bodied men to turn out to assist. "There are no idle men," Reardon reported proudly. "[W]hen one is found he is conscripted to the work of clearing away debris or replanting trees." Members of the Miami Contractors' Association offered to serve destitute homeowners at cost; volunteer crews and Red Cross workers, too, helped make hundreds of dwellings habitable again. In Hialeah alone as many as 112 storm-damaged homes were repaired in a single day, Reardon noted, thanks to a willing army of 600 carpenters, mechanics, and laborers.

Those who could afford to hire help promised workers excellent wages. But in blue-collar Allapattah, tradesmen turned down lucrative job offers until their own neighbors' homes were repaired. Some high-paying work was also rejected as simply too dangerous. On the back of a postcard of the Meyer-Kiser building, for example, one anonymous writer scribbled: "They are paying $4.00 per hour to get this building torn down – only two big iron bolts holding it together. Toombs [*possibly the writer's husband*] said he would not work in it for $10.00 an hour."

*Postcard of the badly-damaged Meyer-Kiser building, predicting it would have to be torn down. In fact, only the top floors of the building were eventually removed. (Dustman Collection).*

With so many helping hands pitching in, progress was swift. Within a week, Miami had "recovered its bearing," Reardon reported, and in another week, "all suffering ha[d] been alleviated by emergency measures." In less than a month, "most of the marks of the storm were gone" – at least according to a hastily-issued pamphlet issued by the Tyler Publishing Company.

Press releases from Miami's mayor E.C. Romfh were decidedly upbeat. "It is the same people who have created the fastest growing city in America who are now turning their energies to the work of reconstruction in Miami," he boasted on September 25.

But even as Miami's rebuilding gained speed and traction, the nation's appetite for images and reports about the devastation was growing. In the worst spirit of yellow journalism, publishers leaped in to capitalize on the tragedy. A pamphlet by the American Autochrome Company of Chicago published shortly after the storm shared one hundred graphic photos of the tragedy, including "many actual scenes of the storm, made while the storm was still raging." The same company also released a smaller publication with forty "ain't it awful" views, a description of the "terrible havoc," and an easy-to-mail accordian folder featuring a map of the ferocious storm's path. And American Autochrome didn't limit itself just to sensational photos of Miami; it also sold photos of the damage in nearby Ft. Lauderdale as well.

Other companies, too, commercialized images of the disaster. H.H. Hamm Company of Toledo, Ohio issued a 25-cent pamphlet featuring 23 photographs of the "Florida Hurricane District." Publisher E.P. Wheelan hawked "Hurricane Scenes: Miami Disaster in Pictures," eye-catchingly subtitled "the most terrific storm in world's history," also for a quarter. Clad in a brilliant red cover, the McMurray Printing Company's "Pictorial History of the Florida Hurricane" shared 47 images of the destruction and five pages of text detailing the tragedy.

Perhaps hoping to cut through at least some of the hyperbole, newspaperman Joe Hugh Reese sat down by candlelight inside his ruined living room on October 5, 1926 to record his personal experiences. Reese's booklet, "Florida's Great Hurricane, Official Account," was a far better-researched account than other hastily-issued compendiums of the day but, like his competitors, featured photos appealing to a sensation-seeking audience.

A plethora of commercial postcards also hit the market. Survivors not only snapped them up as souvenirs of their ordeal but also

*Commercial panoramic photo by Verne O. Williams, showing boats left "high and dry" on Bayshore Boulevard and Royal Palm Park. (Dustman Collection).*

mailed them – along with a few reassuring words – to friends and family across the country. Among the most widely-shared postcards were those by Miami photographer Verne O. Williams, whose crisp black-and-white images and panoramas remain some of the best photographic documentation of the tragic event.

As luck would have it, the hurricane had arrived not long after the debut of affordable personal photography, allowing many survivors to create their own record of the storm. Kodak's first Brownie camera, essentially just a cardboard box with a lens, had first hit the market in 1900, selling for a single dollar. And by the mid-1920s, small box cameras were in widespread use. Various models of the ubiquitous Brownie could be had for between $3 and $6. Specialized cameras able to capture panoramic scenes were also available to both amateur and professional shutterbugs. So by 1926, many households owned cameras. And both during and after the hurricane, they used them.

Untold thousands of images of the hurricane and its aftermath were captured by amateur photographers eager to record the historic moment. They knew they had just survived a life-altering experience, and their cameras allowed them to preserve the evidence. Now-anonymous black-and-white snapshots, relics from long-gone family albums, still turn up for sale on eBay with surprising regularity. This huge body of raw, unfiltered images remains one of the most surprising and enduring legacies of the storm, and a veritable gold mine for historians.

None of the post-storm publicity was good for either local property values or the city's future. With so many negative images circulating via pamphlets and photographs, local boosters kicked their

*A man stands in the doorway of his damaged neighborhood store in South Miami in this amateur image of the hurricane's impact. (Dustman Collection).*

*Another image by the same unknown amateur photographer captured debris at the Miami riverfront. (Dustman Collection).*

own positive public relations efforts into high gear. Bulletins sharing a positive spin on rebuilding efforts were hastily telegraphed to newspapers across the nation, and a makeshift radio station was quickly assembled to help get encouraging words out. The city fathers were determined to send the message that Miami was "down but not out," as newspaperman Reese put it.

Chapter 5: Recovery & Rebuilding

*A small boat "blown ashore" and leaning utility poles were captured in this image of Coconut Grove by an unknown photographer following the storm. (Dustman Collection).*

*Damaged oceanfront estates at 39th Street, sent by someone named Van to a friend named Clifford as part of a Christmas mini-album in December, 1926. (Dustman Collection).*

*A panoramic view of hurricane-damaged South Dade Lumber Company, minus its roof and front wall, viewed from the courthouse looking south. "Took from first rays of dawn till about 9 or 10 o'clock to just get lumber out of streets so cars could go through," the unknown photographer noted on the back. (Dustman Collection).*

These dueling perspectives – alternately highlighting Miami's tragedy or triumph – produced sparks between Florida governor John W. Martin and Red Cross executive director John B. Payne. While Gov. Martin staunchly declared that conditions in Miami were not as bad as reported, Payne protested that Martin was hurting Red Cross fundraising efforts. The Miami Daily News blasted the governor, too, though for the opposite reason: Martin (the newspaper claimed) had allowed Miami to appeal nationwide for aid, but the state had the cash to "take care of her own." Governor Martin, for his part, faced a typical politician's dilemma, unable to make anyone happy. He protested strongly that his statements had been sadly misconstrued. But he also refused to convene a legislative special session to authorize state funds for relief. Happily for hurricane survivors, despite all the mixed messages, relief donations continued to pour in. In just a few weeks, the Red Cross successfully raised over $3.6 million dollars for victims of the storm.

Although the vast majority of hurricane publicity was negative or depressing, a few uplifting and even funny stories emerged in the aftermath. Dredge worker Thomas Gill of Hialeah, for example, had been sound asleep in his bunk when the hurricane hit. The dredge wound up at the bottom of the Bay, and Gill's coworkers were convinced he had gone down with it. Divers later recovered a body from the wreck, and shipmates sadly identified the remains as his. In fact, however, Gill had jumped into the raging waters and managed to swim to shore during the storm. His friends were all gathered to pay their last respects to their comrade and the minister had just finished a reading of the 23rd Psalm when Gill himself strode in, turning his own funeral into a celebration instead.

A second happy story involved a pair who had planned to be married the very day the storm arrived. Ted Yates lived uptown, while his bride-to-be was two miles away at a Hollywood hotel, awaiting the ceremony, when the hurricane struck. Ted tried valiantly to forge his way to her side, but the force of the winds drove him back. When the eye of the hurricane arrived, Ted donned a bathing suit and quickly slogged his way toward the hotel. His fiancee, also clad in bathing attire, rushed out to embrace him. Their joyous reunion was observed by the hotel staff, who somehow managed to procure both a minister and an orchestra in short order. The wedding march was played, "I do's" were exchanged, and the happy couple began their married life in what one report called surf-suits.

Despite the severity of the damage, Miami's businessmen adopted a spirit of deliberate optimism, touting Miami's resilience even before there was any real evidence of that virtue in practice. "[T]he future now lies golden before" Miami and South Florida, newspaperman Reese blithely predicted. "Here is the testing time for Miami, and here is her opportunity," echoed writer Tom Arnold in a *Miami Tribune* editorial. "Up from the ruins of the nation's wrecked playground will come a more beautiful and a more substantial city," predicted the *Miami Daily News* in its own editorial on Monday, September 27, 1926, adding: "The suffering and terror of the catastrophe of a few days ago will become in a brief time only a memory."

Determined to ensure that tourists didn't stay away, the Miami Chamber of Commerce put out a full-color brochure entitled *Miami By the Sea: The Land of Palms and Sunshine*, listing current hotel rates and furnished apartments for rent, and cheerfully proclaiming "We Are Open After the Hurricane." An insert assured would-be visitors that "Miami Conditions [Are] Normal Now."

Perhaps most helpfully of all, a handful of Greater Miami's largest investors made it clear they weren't giving up. Developer Joseph W. Young, who'd built up a fortune from $4,000 invested in the town of Hollywood just five years earlier, blinked tears from his eyes but vowed to build the town back into ten times what it had been. Carl Fisher, one of the major developers of Miami Beach, pledged $2 million dollars toward rebuilding efforts.

As early as September 23, 1926, less than a week after the hurricane hit, survivor L.F. Reardon was reporting an "air of optimism" from managers of the large Miami hotels, none of which had been seriously damaged in the storm. Rather than gloom and doom, he noted, the hotel men were predicting that the coming tourist season would be "Miami's greatest."

The city's fast, comprehensive rebuilding efforts reinforced its sense of invincibility. Miami had bravely stared disaster in the face and was not only coming back, but was "poised once more to prove its claims of being the fairest land on the continent and the playground of the world," as the Tyler Publishing Co. grandly proclaimed in its *Pictorial History*. Miami would bounce back from its frightening and tragic ordeal stronger than ever.

Or so it seemed at first.

But other forces were already coalescing, just over the horizon. Though the Magic City would indeed rebuild swiftly, events outside her control would conspire to sabotage her recovery. And in the end, the return to booming prosperity would take far longer than anyone imagined.

*Broken trees along Brickell Avenue. (Dustman Collection).*

# CHAPTER 6:
# An End and a New Beginning

Mere weeks after the September hurricane, the Magic City seemed to be emerging, phoenix-like, from the devastation. "Florida has a right to hold her head high," proudly proclaimed the *Miami Herald* on October 20, 1926.

Signs of hope and renewal were appearing everywhere. One survivor shared a photo of wind-broken trees along Brickell Avenue, but scrawled encouragingly on the back: "It is like this all the way, but the leaves are coming out green and beautiful."

Boosters had been quick to promise that building back meant building even better. Some even proposed that frame houses be outlawed entirely in Florida, an over-reach that the *Miami Herald* swiftly rebuffed. "The truth is that no type of building had a monopoly on safety in the hurricane, and the frame structures only suffered more because there were far more of them," the newspaper countered in an October 20, 1926 column. Damaged wood-frame buildings "were nearly always of flimsy construction and used for temporary purposes," the newspaper asserted, reminding readers that "very few of [the] old houses were seriously damaged. They were built to withstand storms." The fact that lumber remained a leading industry in the state likely helped torpedo the anti-wood suggestion.

But while recovery efforts were swift, not all parts went smoothly. Insurers found themselves facing losses estimated as high as $7 million dollars, and many homeowners wound up unhappy with the way the companies handled – or *didn't* handle – insurance claims. "Vigorous, and in [some] instances, bitter, complaints are being made of what is termed the 'dilatory tactics' of the insurance companies in adjusting losses," noted the *Miami News* of October 20, 1926. Adding insult to injury, insurers seemed more eager to write new policies – at 50% higher rates – than to settle claims with their existing customers.

But just as Miami's citizens began to relax, confident that the worst of the hurricane's crisis was behind them, Fate showed that it had a few more tricks up its sleeve. Precisely one month and two days after the devastating September blow, a second, smaller hurricane approached Florida's east coast. Forming initially as a low-pressure trough off Costa Rica, this new storm quickly mushroomed to hurricane-force, sweeping across Cuba not once but twice with gusts reaching 150 mph, before spinning northward toward Florida.

Meteorologist Richard W. Gray initially misgauged the new hurricane's trajectory, predicting it would blow itself out somewhere over the Gulf of Mexico. Florida's Seminole Indians knew better. A week before the new storm appeared they began trekking north through Okeechobee toward higher ground, sensing a fresh hurricane was on the way.

They were right. By mid-morning on October 20, Miami newspapers carried warnings about a hurricane of "great intensity" approaching the South Florida coast. By noon, the storm's path had shifted; now it was headed straight at Miami.

Still reeling from the previous catastrophe, frantic residents did what they could to prepare for a second major hurricane. Windows and doors were boarded up, and awnings taken down. Firemen were placed on alert and relief stations readied. Some families abandoned their homes, seeking shelter in buildings they hoped would prove safer. Women and children huddled on the front steps of the Gesu Catholic Church as if hoping for Divine protection, before eventually being herded inside a nearby school.

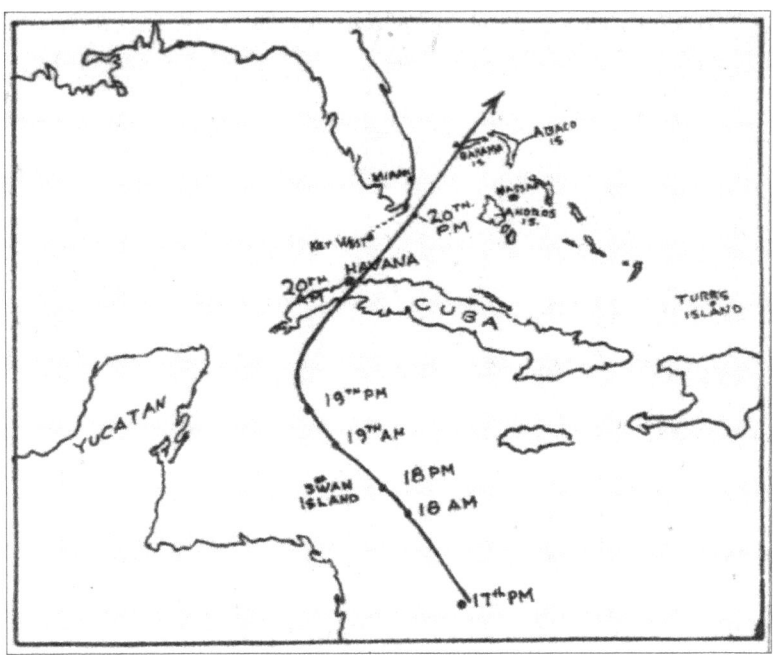

*The trajectory of the October, 1926 hurricane with dates marking its progress as it swung northeast over Cuba and headed for Florida. (Miami News, October 21, 1926).*

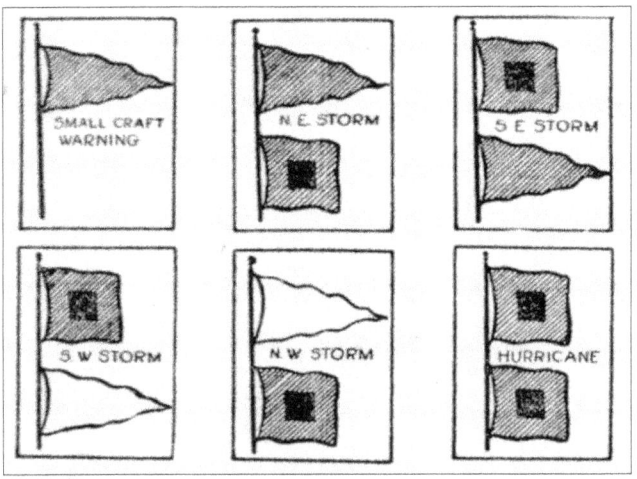

*The Miami Herald published this llustration to help readers interpret storm warning flags.*
*(Miami Herald, October 22, 1926).*

Prisoners were released from Miami's city jail to help their families prepare for the coming storm, based on their solemn promise to return once the danger was over. Members of the Lions Club, deputized a month earlier, were hastily recalled as a "special police corps" to assist Miami Beach police.

By midnight, gale-force winds of 60 to 70 mph had reached Florida's coast. At the last minute, however, the storm's trajectory swerved slightly, sparing Miami the worst of its wallop. Even the damage at Miami Beach proved relatively slight, although fallen trees blocked a section of Collins Avenue temporarily and surf washed portions of Ocean Drive. Four barges were beached, and trolley wires fell. But by and large, Greater Miami escaped serious harm this time.

While this second hurricane proved to be a narrow miss for Miami, Cuba and the Bahamas weren't as lucky. At Havana, a tidal wave breached the Malecon seawall, flooding low-lying areas and destroying over 300 buildings. At least 30 people were killed and as many as 1800 suffered injuries. Forty fishing vessels sank or were damaged, and many smaller craft were swept out to sea. The Bahamas, too, took a heavy beating after the hurricane left Miami and swept east, rendering thousands there homeless.

When blue skies finally dawned again, Miami residents gratefully breathed a sigh of relief. But as hurricane season finally ebbed and gave way to winter in the weeks to come, the city found herself embroiled in a fresh struggle. And this time, the headwinds she faced would be financial ones.

Even before the first hurricane had swept in the previous September, Miami's heady real estate boom of 1920-1923 had already begun to slip away. Land prices had risen too far and too fast, and the speculative bubble in Florida real estate had become all too obvious. Land sharks known as 'binder boys' had been purchasing and then flipping sales contracts as many as eight times in a single day, accompanied by a price mark-up each time.

The pressure on Miami real estate prices had been exacerbated in October 1925 by a railway embargo, as the two major carriers tried to relieve overcrowded rail lines by limiting freight to essential commodities. But this had had the unfortunate side effect of halting the flow of building materials for new construction. A downward skid in real estate prices had begun the following month, driven by material shortages, skittish buyers, unseasonably cold weather, and a growing awareness of just how overheated the real estate market had become.

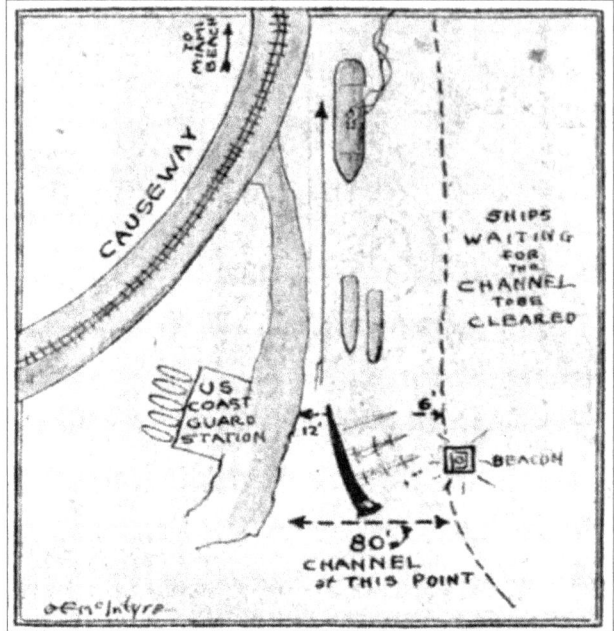

*Illustration showing the wreck of the Prinz Valdemar blocking Miami's 80-foot-wide shipping channel. (Miami News, January 11, 1926)*

A shipping fiasco in Miami's harbor in early 1926 only compounded the city's woes. On January 10, 1926, the *Prinz Valdemar*, a 241-foot barkentine, capsized in the turning basin at the mouth of Miami's harbor, wedging herself sideways in the ship channel known as Government Cut. Between her submerged steel hull and her four tall masts, the wreck effectively blocked both incoming and outgoing vessels. Some forty ships were suddenly trapped in port, while sixty incoming vessels were forced to loiter outside the harbor, unable to reach the docks to discharge their cargo.

Commercial shippers found themselves losing thousands of dollars per day. Even more important for the beleaguered city, construction supplies were unable to reach desperate builders for nearly a month. At long last, twenty-five long days after the *Prinz Valdemar*'s capsize, the channel was finally reopened to shipping traffic on February 4, 1926.*

Visitor traffic in early 1927 remained understandably slack, following the two separate hurricanes and growing national concern over Florida inflated real estate. But local promoters did their best to put a positive spin on things. In early April, the *Miami Herald* reported that a "record-setting" crowd had descended on Miami Beach. Its competitor, the *Miami News*, also did its best to help bolser Florida's reputation, quoting an apocryphal visitor as swearing, "I would rather be blown away to my home in the starry heavens on a Florida hurricane than to die on a feather bed in any other state in the union."

> SOUTHERN PINE—THE SUPREME STRUCTURAL WOOD OF THE WORLD
>
> **HURRICANE PROOF CONSTRUCTION**
> SEE
> **The Miami Model House**
> TODAY
> AT
> GRAPELAND BOULEVARD
> (27th AVE.)
> AND
> S. W. SIXTH STREET
> BEACOM MANOR
> ON DEMONSTRATION DURING THE PERIOD OF
> ACTUAL BUILDING OPERATIONS

*An ad for "hurricane proof" pine homes. (Miami News, February 27, 1927).*

But could Florida continue to attract the desperately-needed winter visitors upon which its economy had grown so dependent? Ah, that was an open question, and one to which newspapers and local boosters alike paid considerable attention.

Miami pioneer Isidor Cohen did his best to inspire confidence with vigorous jawboning, boldly predicted that the coming 1927-28 winter season would be Miami's greatest ever. He claimed to have witnessed a "decided change in favor of Miami" among New York acquaintances, undeterred by "disappointed speculators and envious detractors." State Hotel Commissioner Jerry W. Carter came up with a creative state-wide marketing plan offering early winter and

---

*\* Though the Valdemar was successfully refloated, she would prove a navigational thorn in Miami's side in years to come. The derelict ship was initially towed to the Municipal piers, then moved to an anchorage spot opposite S.E. First Street. But when the September 1926 hurricane hit, she broke her loose from her moorings and ran aground again, this time in the Cape Florida channel.*

*A flurry of buck-passing ensued for over a year. County commissioners washed their hands of the matter, claiming no responsibility for the wreck. The pilot commission acknowledged it had jurisdiction to remove it, but argued they lacked funds to do so. By November 1927, the once-proud Valdemar was "wallowing on her side like a dyspeptic sea cow in front of Bayfront park," as the Miami Herald put it – and still a menace to navigation. She was eventually moved to the north end of the city yacht basin at the foot of N.E. Sixth Street and corralled inside a steel bulkhead. After weeks of frenetic remodeling, she was reborn on May 1, 1928 as an aquarium, a tourist attraction operated by Capt. R.J. Walters.*

late spring visitors a total of 40 nights of lodging and discounted attractions, all at a bargain price, in hopes that inducement would lengthen the usual tourist season.

In fact, however, many previous Florida visitors were *afraid* to return. A report prepared by the Florida State Hotel Association confirmed this reluctance had nothing to do with avoiding hurricanes and everything to do with escaping the clutches of the local courts. Those who had defaulted on real estate contracts signed during Florida's excitement were eager to distance themselves from those unpaid debts. As a result, the report advised, tour companies were increasingly booking travel to California for their clients, instead of arranging trips to Florida.

One humorous observer would later say, "About the only thing that kept the tourist season [of 1927-28] from being a complete flop was the supply of good whiskey." Prohibition may still have been technically in force, but economics was economics. Visitors to Florida liked to drink, so enforcement tended to be lax. At least Miami had *that* going for it.

And then, as if Mother Nature herself was holding a grudge, a *third* hurricane struck South Florida on September 17, 1928 – almost two years to the day after Miami's original 1926 disaster. What became known as the Okeechobee Hurricane, like the second one in October 1926, didn't strike Miami directly, making landfall instead near West Palm Beach. But it would prove to be the third-deadliest in United States history. At least 2,500 Floridians drowned, and more than 1,700 homes were destroyed. Flood waters reached twenty feet deep in some places.

The news of not just one but *three* South Florida hurricanes within such a short span wreaked predictable havoc with Miami property values. As noted Miami historian Seth Bramson has noted, just as Miami led the rest of America into the booming 1920s, so, too, did these triple disasters propel the city into an early decline. "In effect," observed Bramson, "Miami experienced the Great Depression three years earlier than the rest of the country."

Immediately before this third hurricane put in its most unwelcome appearance, both tourism and real estate sales had experienced a slight but welcome up-tick. Even after the October, 1928 storm had passed, there was still reason for hopes to remain high. The 264-mile-long Tamiami Trail linking Tampa to Miami had just been completed the previous April, making Miami more accessible to tourists than ever. The lavish 28-story Miami-Dade Courthouse

*Wire service aerial photo of Miami's harbor and skyscrapers, touting the city as a "Little New York." The city's future would change dramatically less than four months later with the stock market crash. (P&A Photos, July 2, 1929).*

at 73 W. Flagler Street, a stunning $4 million dollar project begun in 1926, had finally been finished and a joyful public ceremony had just dedicated the building on September 6, 1928. The Flagler System's iconic but heavily-damaged Royal Palm Hotel had been successfully repaired and stood ready to welcome visitors for the 1928-1929 winter season.

But in October 1929, the Roaring Twenties came to a chaotic and tumultuous end. The stock market crashed, and soon the entire nation found itself in the painful grip of the Great Depression.

Florida's visitor numbers collapsed, dropping precipitously from three million per year to just one million. Miami's tourist-based economy cratered. Construction activity plummeted. Unemployment skyrocketed.

Miami was no longer a 'playground for the rich.' Fat wallets were out. Penny-pinching was in. As the nation adjusted its spending habits to new economic constraints, the hospitality industry was forced to shift its focus from the affluent traveler to those seeking moderately-priced hotels or even 'tin can tourist camps.'

Chapter 6: An End and a New Beginning

*A Florida 'tin can tourist camp,' from a 1930s postcard. (Dustman Collection).*

*Photo of the aging Royal Palm Hotel in 1930, captioned by the unknown photographer: "Last picture of the Royal Palm Hotel before wreckers started to work." (Dustman Collection).*

*A 1908 postcard showing the iconic Royal Palm Hotel in her hey-day. (Dustman Collection).*

Miami's once-lavish grand hotels suffered the most as visitor tastes and budgets changed. The still-gracious but aging Royal Palm became one of the downturn's first victims. When the Flagler corporation discovered the elderly hotel was infested with termites, it made the painful but practical decision to simply tear it down. The iconic structure was demolished in June 1930.

Despite Franklin D. Roosevelt's inaugural address in 1933 assuring Americans they had nothing to fear but "fear itself," the Great Depression ground on through the rest of the decade. The economic crunch eroded not only Miami's fortunes but also the wealth of some of its most prominent citizens. Pioneering Miami Beach developer Carl Fisher, whose net worth had topped fifty million dollars in 1925, found himself 'dead broke' by the late 1930s. Fisher would pass away July 15, 1939 of gastric hemorrhage, a condition that may have been linked to his alcoholism.

Ironically enough, it would take yet another catastrophe – the Second World War – to finally return Miami to prosperity. War-time manufacturing and military training bases boosted local employment. And once the war was over, returning GIs flocked to Florida as visitors and residents, remembering its balmy weather.

Today, the 1926 hurricane remains a largely-unremembered turning point in Miami's fabled history. Those who lived through it, like my father, certainly never forgot. But the event itself is now a lifetime away. The people who actually experienced it have turned to dust, and those first-hand memories are gone. Meteorologists still acknowledge the storm's horrific magnitude. But with its fury one hundred years in the rear-view mirror, the "Great Hurricane" rarely gets mentioned. Other more recent hurricanes have seized the attention of younger generations.

As terrible as the 1926 hurricane was, Miami's sheer tenacity prevailed in the end. She not only endured all that Mother Nature threw her way, but found ways to soldier on. Improved building codes were developed; better disaster preparedness became mainstream. New rebuilding opportunities were seized, and fresh economic drivers embraced. As the next one hundred years slipped by, Miami did more than simply reclaim her place in the sun. She found new ways to thrive.

Still, there was a reason that the hurricane of September 18, 1926 was long spoken of as the 'Great Hurricane.' That terrifying, awful, sometimes-miraculous, often-tragic day marked a turning point in Miami's history, the unexpected blow that burst Miami's real estate bubble and propelled her headlong into the Depression.

It was a day that altered Miami's future – and the lives of so many of her citizens. May those who lived through it, those who perished, and the tale of the Great Hurricane itself all be long remembered.

## Bibliography & Resources

### Books:

The Florida Hurricane & Disaster: 1926-1992, Howard Kleinberg, L.F. Reardon (Centennial Press 1992).

### Magazines, Pamphlets, & Other Publications:

"23 Views Florida Hurricane District, September 18th, 1926" (H.H. Hamm Co., Toledo, OH).

"30 Views of Hurricane at Fort Lauderdale, Sept. 18, 1926" (postcard folder published by American Autochrome Co., Chicago).

"40 Views of Miami Hurricane, September 17-18, 1926 With Full Description of Its Terrible Havoc" (American Autochrome Co., Chicago, 1926).

"100 Views, Miami Hurricane, With Many Actual Scenes of the Storm; The Florida Hurricane Which Devastated Miami, Hollywood, Fort Lauderdale, Palm Beach, Moore Haven, and Other Towns and Cities on September 18th, 1926" (American Autochrome Co., Chicago, 1926).

"A Pictorial History of the Florida Hurricane, Forty-Seven Views and Five Pages of Information, September 18, 1926" (McMurray Ptg. Co., Miami, Florida, 1926).

"Actual Scenes of Miami Hurricane of Sept. 18, 1926, Made While Storm Was Raging and on Day of Storm" (postcard folder published by American Autochrome Co., Chicago).

"Along the Florida Reef," *Harper's Monthly* (February, March, and April 1871).

"Florida's Great Hurricane, Official Account," by Joe Hugh Reese (Lysle E. Fesler, publisher, 1926).

"Hurricane Scenes: Miami Disaster In Picture[s] (McCluney Ptg. Co., Inc., 1926).

"Illustrated Story of the Hurricane That Wrecked the Miami District, on the East Coast of Florida, September 17th and 18th, 1926," by J. Wadsworth Travers (The Palm Beach Press, 1926).

"Luten Bridge in Tropical Hurricane" (Luten Bridge Co., 1926).

"Miami the Beautiful" (published by Foster & Reynolds Co., Miami, Florida, *circa* 1920).

"Personal Glimpses: Human Flotsam of the Florida Hurricane," *Literary Digest* (October 9, 1926)(eyewitness accounts).

"Souvenir of Miami, Florida, The Magic City" (C.T. & Co., *circa* 1923).

"Update," Historical Society of Southern Florida (Vol. 8, No. 3, August 1981, "Letter to Mother" by George Herbert Cooper); (Vol. 5, No. 6, August-October 1978, "And I Knew Hurricanes," by Mispah Otto de Boe)(personal accounts).

### Newspapers (from HistoryMiami Archives):

*Coconut Grove Times*, (Vol. 4, No. 31), September 24, 1926.

*Miami Daily News*, September 18 and 19, 1926.

*Miami Herald*, September 21, 1926.

Additional issues from September/October, 1926 available through *Newspapers.com*.

### Other Research Sources:

Archives of the Historical Museum of South Florida (HistoryMiami), 101C West Flagler St., Miami FL 33130 (first-person accounts and letters recalling the storm).

## About the Author

A former lawyer and prosecutor, Karen Dustman layered in a separate career as a journalist. Over 200 of her freelance articles have appeared in national and regional publications, and she continues to wrangle an active history blog. Through her indie imprint, *Clairitage Press*, she's published two dozen previous history books, including <u>Bad Boys Miami</u> (vintage true crime stories from Miami's past) and <u>Dr James M. Jackson: Miami's Beloved First Doctor</u>.

Despite growing up in a historic seafaring town, Karen detested history in school. "They always focused on the boring stuff – names, dates, generals," she chuckles. Today, it's the people-stories she loves sharing most. "The struggles. The losses. Finding a rich gold seam and then losing everything in a poker game. It's not the bare facts and figures that call to us," she says. "It's the people who lived an incredible life and how they managed to survive that's fascinating."

This book is in part a tribute to Karen's father, a native of Arch Creek, Florida, and a survivor of the 1926 Miami hurricane.

For more of Karen's books, visit *www.Clairitage.com*, or find her on *Amazon.com*.

www.ingramcontent.com/pod-product-compliance
Lightning Source LLC
Chambersburg PA
CBHW062103290426
44110CB00022B/2694